Inorganic Reactions in Organized Media

Inorganic Reactions in Organized Media

Smith L. Holt, EDITOR
Oklahoma State University

Based on a symposium sponsored by the Division of Inorganic Chemistry at the 181st ACS National Meeting, Atlanta, Georgia, March 29–April 3, 1981.

ACS SYMPOSIUM SERIES 177

AMERICAN CHEMICAL SOCIETY
WASHINGTON, D. C. 1982

6554-2435
CHEM

Library of Congress CIP Data

Inorganic reactions in organized media.
(ACS symposium series, ISSN 0097-6156; 177)

"Based on a symposium sponsored by the Division of Inorganic Chemistry at the 181st ACS National Meeting, Atlanta, Georgia, March 29–April 3, 1981."
Includes bibliographies and index.

1. Chemical reaction, Conditions and laws of—Congresses. 2. Chemistry, Inorganic—Congresses. 3. Micelles—Congresses. 4. Emulsions—Congresses.
I. Holt, Smith L., 1938– . II. American Chemical Society. Division of Inorganic Chemistry. III. Series.

QD501.I624 541.3'94 81-20626
ISBN 0-8412-0670-8 ACSMC8 177 1-254
1982

PRINTED IN THE UNITED STATES OF AMERICA

ACS Symposium Series

M. Joan Comstock, *Series Editor*

FOREWORD

The ACS Symposium Series was founded in 1974 to provide a medium for publishing symposia quickly in book form. The format of the Series parallels that of the continuing Advances in Chemistry Series except that in order to save time the papers are not typeset but are reproduced as they are submitted by the authors in camera-ready form. Papers are reviewed under the supervision of the Editors with the assistance of the Series Advisory Board and are selected to maintain the integrity of the symposia; however, verbatim reproductions of previously published papers are not accepted. Both reviews and reports of research are acceptable since symposia may embrace both types of presentation.

CONTENTS

PREFACE

The symposium, "Inorganic Reactions in Organized Media," and this volume grew out of the editor's belief that, while organic chemists have paid due attention to reactions in organized media, (that is, micelles, microemulsions, and vesicles), inorganic chemists as a group have virtually ignored the potential utility of these systems. Indeed, if one surveys the literature, he/she will be struck by the innumerable uses that have been found by the organic and physical chemists for organized media, yet how paltry are the references to reactions that involve inorganic species. As a consequence, it was deemed desirable to bring together not only inorganic chemists who are working in structured solutions, but also those individuals in other disciplines who have acknowledged expertise in the area. In this way, new insight may be developed into ways in which organized media can beneficially be used in the study of inorganic reactions and mechanisms.

A variety of interactions are examined in this volume. The first paper provides a general background on structures obtained from surfactant association and references some of the important literature in the area. The next five chapters focus on photochemical processes in organized assemblies. Chapter 2 focuses on cage and magnetic isotope effects of micellization. Chapter 3 discusses light-induced electron-transport processes in micellar systems and Chapter 4 addresses photoprocesses in synthetic vesicles. Chapter 5, 6, and 7 describe exciting new developments in the study of photoinduced reactions at the surface of colloidal oxides. Of particular importance are studies involving the photochemical splitting of water and hydrogen sulfide discussed in Chapter 7. Chapter 8 describes work wherein micelle-like structures, which modify electrochemical reactivity, are formed on platinum electrodes when surfactant is introduced into particular systems.

Chapters 9 through 12 focus on the use of microemulsions to modify reaction rate and pathway. Included are studies of porphyrin metalation, transmetalation, the Wacker process, and the hydrolysis of chlorophyll.

The last two chapters deal with chemical reactions in solids and solid surfaces, bridging the final gap between reactions in organized liquids and reactions in crystalline solids.

SMITH L. HOLT
Dean, College of Arts and Science
Oklahoma State University
Stillwater, OK 74074

September 1, 1981

1

Surfactant Association Structures

STIG E. FRIBERG and TONY FLAIM

University of Missouri—Rolla, Department of Chemistry, Rolla, MO 65401

The phase regions for micellar solutions and lyotropic liquid crystals form a complicated pattern in water/amphiphile/hydrocarbon systems. The present treatment emphasizes the fact that they may be considered as parts of a continuous solubility region similar to the one for water/short chain amphiphilic systems such as water/ethanol/ethyl acetate.

Hence the different phases may be visualized as a series of association structures with increasing complexity from the monomeric to the liquid crystalline state. The transfer from the monomeric state to the inverse micellar structure is discussed for two special cases and it is shown that packing constraints may prevent the formation of inverse micelles. Instead a liquid crystalline phase may form.

The surfactant association structures have a long history of research ranging from the McBain introduction of the aqueous micellar concept[1] over the interpretation of micellization as a critical phenomenon[2,3] to the analysis of the structure of lyotropic liquid crystals[4] and the comprehensive picture of the phase relations in water/surfactant/amphiphile systems.[5] These studies have emphasized the relation between the association structures in isotropic liquid solutions and the liquid crystalline phases. Parallel extensive investigations in crystalline/liquid crystalline lipid structures[6,7] have provided important insight in the mechanisms of the associations.

The thermodynamics of these systems have been extensively discussed in recent years including the micellization,[8,9] the liquid crystals[10] and inverse micellization.[11,12] In addition the more general problem of the stability of microemulsions has been extensively covered.[13-16]

0097-6156/82/0177-0001$05.00/0

This article will, in addition to a short description of the essential features of surfactant systems in general, concentrate on the energy conditions in premicellar aggregates, the transition premicellar aggregates/inverse micellar structures and the direct transition premicellar aggregates/lyotropic liquid crystals.

Surfactant Systems - A Word of Caution

The inverse micellar solubility areas in systems of water, surfactants and a third amphiphilic substance frequently are of a shape according to Fig. 1.[17] Such shapes are also found in W/O microemulsions[18,19] when water solubility is plotted against cosurfactant/surfactant fraction.

It is tempting to evaluate this solubility curve as showing a maximum of water solubility at the apex point.

It is essential to realize that any thermodynamic evaluation of this solubility "maximum" with standard reference conditions in the form of the three pure components in liquid form is a futile exercise. The complete phase diagram, Fig. 2, shows the "maximum" of the solubility area to mark only a change in the structure of the phase in equilibrium with the solubility region. The maximum of the solubility is a reflection of the fact that the water as equilibrium body is replaced by a lamellar liquid crystalline phase. Since this phase transition obviously is more related to packing constraints[8] than enthalpy of formation[20] a view of the different phases as one continuous region such as in the short chain compounds water/ethanol/ethyl acetate, Fig. 3, is realistic. The three phases in the complete diagram, Fig. 2, may be perceived as a continuous solubility area with different packing conditions in different parts (Fig. 4).

This means that the phase changes observed have comparatively less importance for the thermodynamics of the system. On the other hand, the changes and modifications of the association structures <u>within</u> the isotropic liquid hydrocarbon or alcohol phase pose a series of interesting problems. Some of these have recently been treated in review articles by Fendler[21] who focussed on surfactant inter-association emphasizing consecutive equilibria and their thermodynamics. The following description will focus on the intermolecular interaction between different kinds of molecules and the importance of these interactions for the "inverse" association structures.

It should be emphasized that these structural changes within a one-phase region may change the kinetics of a chemical reaction in a pronounced manner. As an example may be mentioned the catalytic effect of inverse micelles on ester hydrolysis. Fig. 5 is from the first publication[22] on this subject. It clearly shows the lack of catalytic effect by the premicellar aggregates and the sudden increase of hydrolysis rate in the concentration range where the inverse micelles begin being formed.

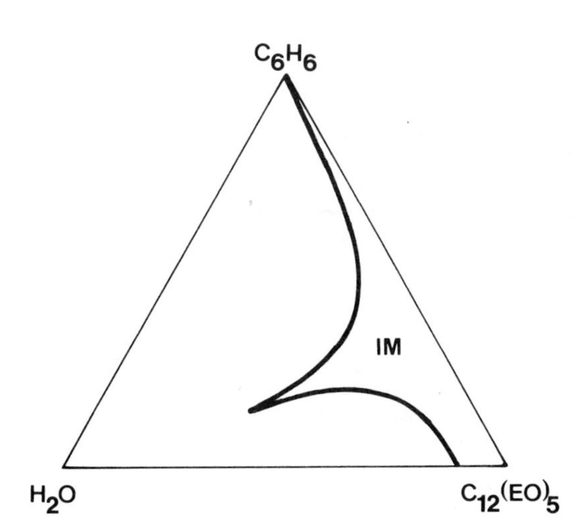

Figure 1. The solubility area of water in a hydrocarbon (C_6H_6) penta(ethylene glycol) dodecyl ether ($C_{12}[EO]_5$) solution at 30°C. Key: IM, inverse micellar solution.

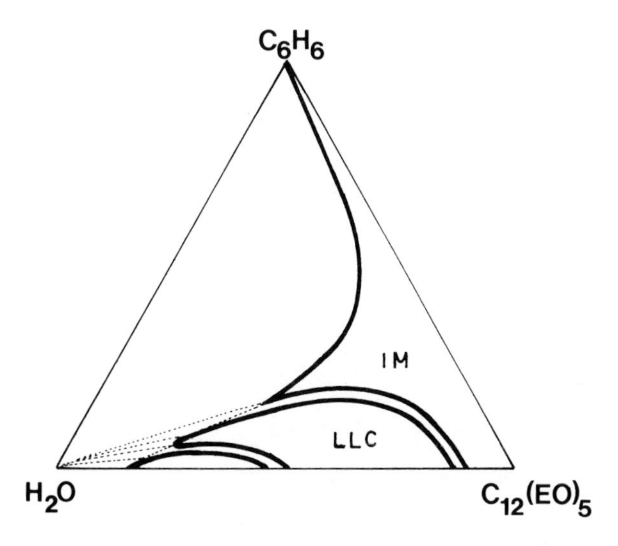

Figure 2. The phase diagram water/benzene/penta(ethylene glycol) dodecyl ether at 30°C. Key: IM, inverse micellar solution; LLC, lamellar liquid crystal; and unmarked, aqueous micellar solution.

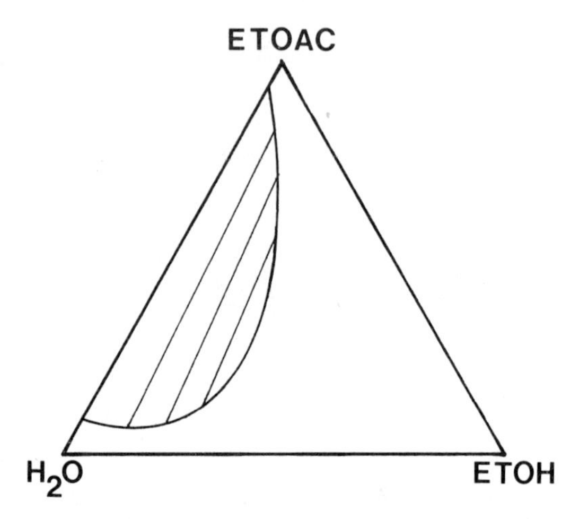

Figure 3. The phase diagram water/ethanol (ETOH)/ethyl acetate (ETOAC).

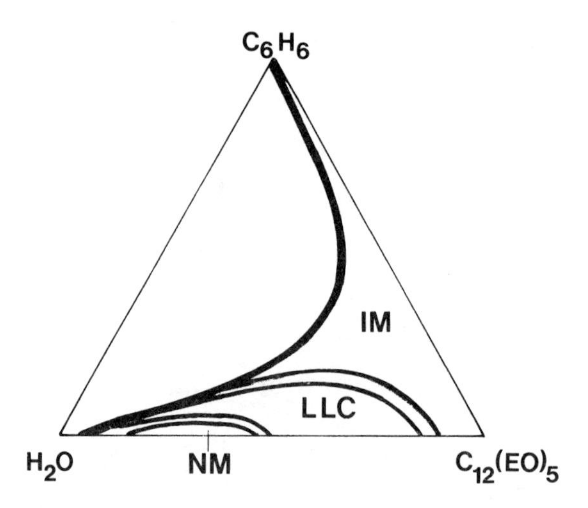

Figure 4. The similarity of Figure 3 to Figure 4 is seen if the maximum solubility of water is emphasized.

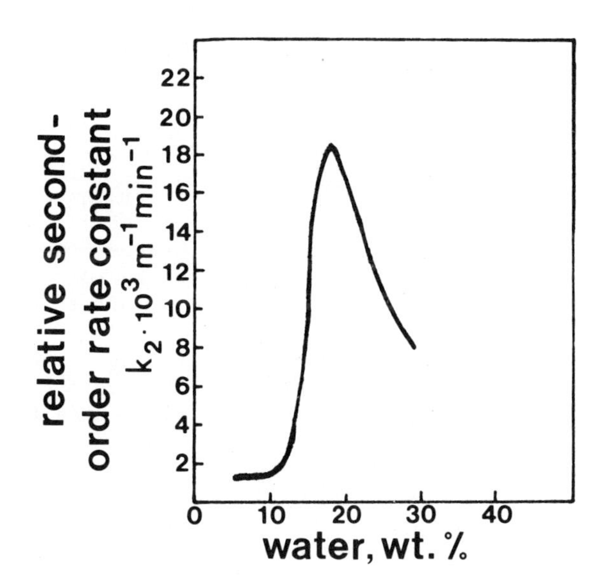

Figure 5. The premicellar aggregates (< 12% water) do not catalyze the hydrolysis of an ester (22).

Two solutions to illustrate these structural changes within one phase were chosen with emphasis both on their practical importance and on the pronounced difference in behavior brought about by a small difference in composition. The two liquids chosen are both soap/water solutions in a) an alcohol and b) a carboxylic acid. The alcohol solutions form the basis for the W/O microemulsions[(18)] and the carboxylic acid solutions show a pronounced difference in properties which merit an evaluation of the intermolecular forces between the solvent and the solute. The primary difference is in the minimum concentration of water that is required to form the solution.

Minimum Water Content

The two inverse micellar solutions will first briefly be described followed by an evaluation of the available experimental and theoretical information.

The Sodium Carboxylate/Carboxylic Acid System. This system has been investigated using IR and NMR[(23)] showing the extremely strong hydrogen bonds present. It appeared evident early that the strong hydrogen bonds between ionized and nonionized carboxylic groups were responsible for the stability of the 4 acid:2 soap complex[(24)] but no information could be obtained on the energy for formation of the complex nor could the stability be based on a thermodynamic calculation.

An answer has been given after calculations of the binding energies using the CNDO/2 semi-empirical approach[(25)] to quantum mechanical calulations. The results[(26)] illustrated the importance both of the hydrogen bond and the carbonyl/sodium ion ligand bond of the proposed structure. The values for the hydrogen bond strength (19.5 Kcal/mole, 82 kJ/mole) compared well with the generally accepted values for other ion/molecule systems.[(28)] In addition to the hydrogen bond the 4:2 molecular compound is stabilized by the acid carbonyl/sodium ligand bond, 36.1 kcal/mole (151 kJ/mole).

The calculations[(27)] involving a thermodynamic cycle from the solid crystalline soap and the liquid acid as references states showed the 4:2 complex to be stable even in the gaseous state. The transition from gaseous to liquid state should give further stabilization to the structure. The excellent stability of the compound was illustrated by the fact that dilutions to a composition of the 7% sodium octanoate/octanoic acid, 93% by weight CCL_4, gave no change of the structure.

These conditions are conspicuously in contrast with those in the carboxylate/alcohol system.

The Carboxylate/Alcohol System. A comparison between different alcohol solubility areas with different surfactants and water[(5)] reveals the fact that a minimum water/surfactant ratio is necessary in order to ensure solubility of the surfactant in the alcohol. This is in contrast to the behavior of a liquid carboxylic acid that will dissolve a soap with no water present[29] (Fig. 6).

The difference is pronounced. In an alcohol solution a minimum of approximately six water molecules are required per soap to bring it into solution. A liquid carboxylic acid will dissolve the soap without water to a soap/acid molecular ratio of 1/2. It appears reasonable to evaluate these differences from terms of intermolecular forces. These forces, the strong hydrogen bonds and ligand bonds to the metal ion will be treated in the following section.

Experimental Information. The review by Ekwall[5] offers a whole series of phase diagrams which all show similar behavior. In order to dissolve an anionic surfactant with a sodium counter ion in an alcohol a minimum water/surfactant molar ratio of about six is needed to achieve solubility. The corresponding ratio for the potassium ion is three.

At the same time investigations using light scattering, electron microscopy, positron annihilation, dielectricity and transport properties[19,30-34] indicated the surfactant molecules not to be involved in associations to colloidal size aggregates at these low water contents. The low light scattering intensity rather points to the surfactant molecules not to be inter-associated (Fig. 7).

An approximate thermodynamic evaluation[27] based on liquid water and crystalline sodium octanoate as reference states has recently evaluated the energy conditions in the premicellar aggregate. The calculations essentially were concerned with the free energy of a gaseous soap/water complex. No attempt was made to evaluate the chemical potential changes in any of the components when dissolved in the pentanol.

Accepting the facts that the calculations may be considered a zero order approach, the results are illustrative of the reason for the stability of these small aggregates. The calculations included heat of evaporation of the surfactant, heat of evaporation of water and the free energy association in the gaseous phase. The enthalpy of the latter was calculated using the CDNO/2 approach[25] modified for larger aggregates.[35]

CDNO/2 has a proven record for accuracy in describing the energy of interaction for water clusters around metal ions in aqueous solutions.[36,37] The entropic contribution due to the different spatial arrangements of the particles was calculated using the liquid volume according to Ruckenstein[13] and Reiss.[38]

The following model was employed. The water molecules were consecutively entered into a regular octahedron around the sodium ion with the two first water molecules hydrogen bonded to the carboxylate group. Water molecules in excess of the six first

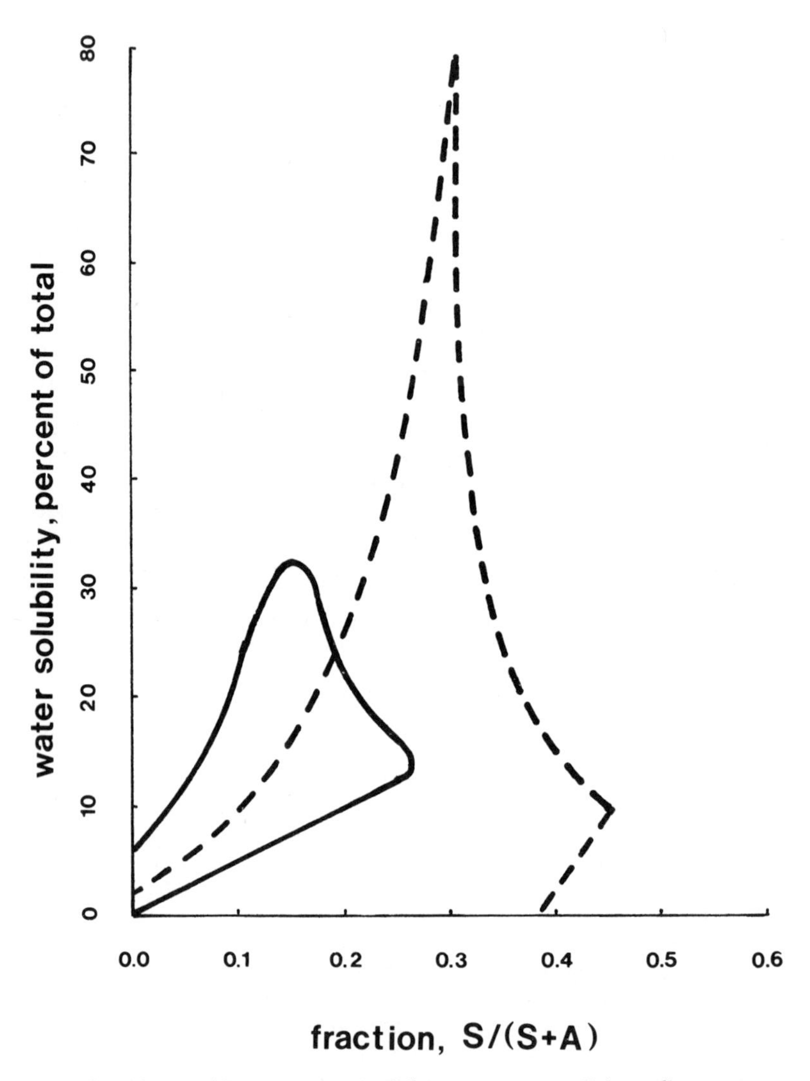

Figure 6. The solubility of water (weight percent of total) in sodium octanoate (S)/octanoic acid (A) (– – –) and sodium octanoate (S)/octanol (A) (——) mixtures (weight fraction).

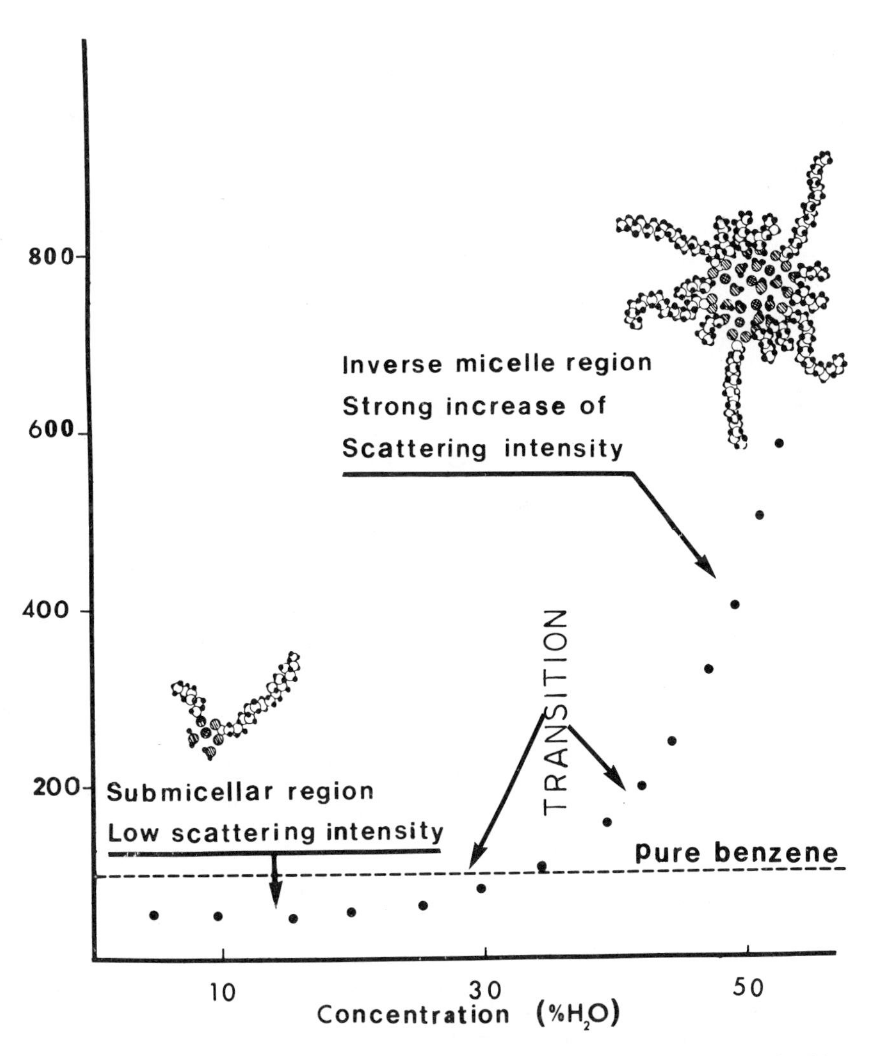

Figure 7. At low water content (< 35% by weight) the water is not involved in colloidal association structures in a pentanol/potassium oleate (weight ratio 3.0) solution.

ones were hydrogen bonded forming an outer shell to the primary octahedron. The results are given in Fig. 8.[27]

It is obvious that the results depend on the specific model chosen and that greatest care should be exercised when conclusions are drawn. However, the following remarks about the results may be justified.

The high hydration energy for the first added water molecules are primarily responsible for the thermodynamic stability of these premicellar aggregates. The calculations show a minimum of five water molecules to be necessary for the stability of the gaseous complex. The agreement with experimental values, 5.5-6 H_2O/soap,[5] is excellent. This coincidence is to some extent fortuitous, but the strong negative shape of the free energy curve (Fig. 8) substantiates the claim of the main driving force being the high heat of hydration for the initially bound water molecules.

Experimental results[19,30-34] show a maximum of ten water molecules per soap molecule to be stable for this monomeric specimen. According to the calculations and to the chosen model the stability ends at thirteen water molecules per soap molecule giving a positive free energy and instability for this monomeric specimen at higher water contents. Again the general trend supports the role of a large hydration energy as the stabilizing factor.

Stable structures at high water contents could be liquid water, spherical inverse micelles or liquid crystals with lamellar or "inverse hexagonal" structure. The transition to the two last structures will be discussed in the next section.

Premicellar Aggregates/Inverse Micelles/Liquid Crystals

The calculations presented in the preceeding section agree well with experimental evidence. The light scattering studies of the system water/potassium oleate/pentanol[19] show a transition from premicellar aggregates to inverse micelles at a water/soap molecular ratio of 10 and an alcohol/soap molecular ratio of ten and higher. The series with the higher soap content, alcohol/soap molecular ratio 5.5, did not show micellization; instead a lamellar liquid crystal was formed when association in excess of the premicellar aggregates took place. This phenomenon can be understood in a qualitative manner using semi-empirical thermodynamics.[8,39]

The chemical potential of an associated amphiphile, μ_n^o, is taken as[8]

$$\mu_n^o = \gamma \cdot a + \frac{2\pi e^2 D}{E_a} + g$$

in which γ is the interfacial tension at the interface of the association structure, a is the amphiphile cross-sectional area

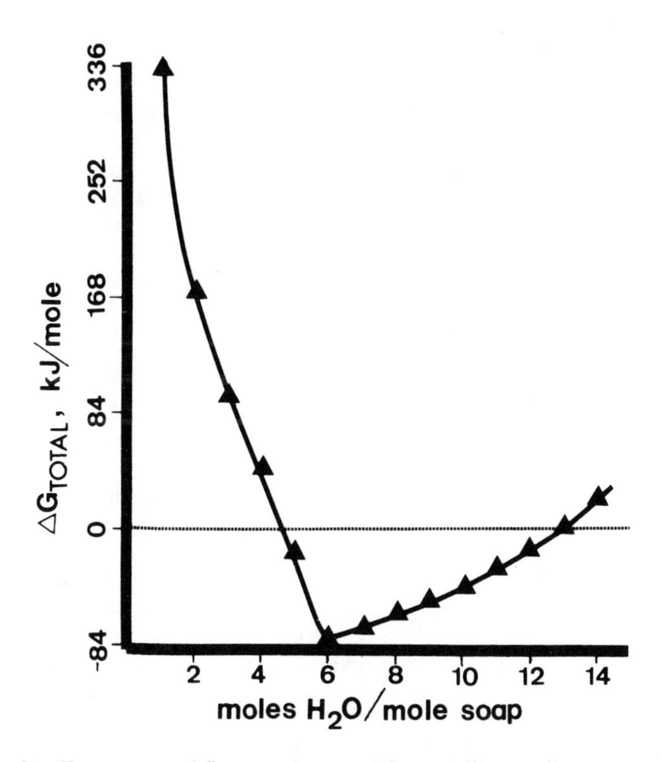

Figure 8. Free-energy difference from solid crystalline sodium octanoate and liquid water for a (mono) sodium octanoate/water molecular compound with different numbers of water molecules.

being in contact with the water, e the electronic charge, D is the Debye distance, E_a the dielectric constant and g the energy to transfer the hydrocarbon chain from water to the hydrophobic part of the association structure.

This approach can be used to give critical micellization concentrations as well as size distributions of aggregates[8] but is, of course, unable to distinguish between different forms of the association structures. In order to achieve this packing, restraints must be introduced.[8]

The relevant term is the expression $\frac{v}{a_o l_c}$ in which v is the volume of the hydrocarbon chain, a_o is the cross-sectional area of the amphiphile (assumed constant) and l_c is the critical chain length (≃ 90% of the fully extended hydrocarbon chain). With these postulates[8] the following conditions may be stated (Table I).

The original contribution[8] applied these conditions to two-phase systems of amphiphile and water. An extension to a three-component system encounters partitional problems, since the exact location of the third molecule may not be known.

The information about area per head-group available is from low angle X-ray data on liquid crystals. Ekwall[5] early found that the mean area in a binary system obeys the relation

$$S = S_1 Z^g \qquad \text{or}$$

$$\log S = \log S_1 + g \ln Z$$

in which S is area per head-group, S_1 is area per head-group with 1 H_2O, Z is number of water molecules per soap and g is a constant ≃ 0.2.

For soap/alcohol combinations[5] g will depend not only on the soap counter ion but also on the alcohol/soap ratio. Furthermore, when a certain alcohol/soap ratio is exceeded (≃2 for the potassium oleate system) S becomes independent of the water content of the lamellar phase. This condition applies for inverse structures and the water/pentanol/potassium oleate inverse micellar system will be examined for the structure determining ratio in Table I.

The mean polar head group area in a water/decanol/potassium oleate system is approximately 26 $Å^2$ for an alcohol/soap molecular ratio of 2.1[5]. For the soap alone an area of 36.0 $Å^2$ was found. Assuming a linear relationship a value of 22 $Å^2$ is obtained for the decanol. This value will be used also for the pentanol. The volume, chain length and area can now be estimated[8]

$$v = 27.4 + 26.9 \frac{R_a + 17}{R_a + 1} \quad (Å^3)$$

$$l_c = 0.9 \left(1.5 + 1.265 \frac{R_a + 17}{R_a + 1}\right) \quad (Å^3)$$

TABLE I.

Conditions for different amphiphilic association structures[36]

$\frac{v}{a_o l_c}$	Preferred structure
$\leq 1/3$	Aqueous, spheric micelles
$\geq \frac{1}{3}, \leq \frac{1}{2}$	Cylinders with polar groups outwards
$\geq \frac{1}{2}, \leq 1$	Bilayers
≥ 1	Inverse structures

$$a_o = \frac{36 + 21\ R_a}{1 + R_a}$$

and the $\frac{v}{a_o l_c}$ can be calculated for an inverse micellar solution.

The experimental results for the water/pentanol/potassium oleate system (18,19) show that a pentanol/potassium oleate molecular ratio of 5.5 and lower should give a premicellar aggregate/lamellar liquid crystal transition instead of the premicellar aggregate/inverse micelle transition at high alcohol/soap ratios.

Calculation of the expression $v/a_o l_c$ for a pentanol/potassium oleate ratio of 5.5 gives the result

$$v/a_o l_c = 0.98$$

A pentanol/potassium oleate ratio of 15 that is typical of the inverse micellar solution gives the corresponding value 1.02. Formally the two values are straddling the value $v/a_o l_c = 1$ in the correct directions, but it is obvious that they are extremely similar and the application of the zeroth order approach[8] to these systems must be viewed with caution. The pronounced influence of a partition of cosurfactants between the interface and the organic bulk is evident.

Summary

It has been shown that the presence of strong intermolecular forces may have a drastic effect on the stability of premicellar aggregates in inverse micellar systems.

The stability of a carboxylic acid/soap premicellar aggregate was shown to be due to the presence of extremely strong hydrogen and ligand bonds. In alcohol systems the corresponding stabilizing energy was the large heat of hydration of the water molecules forming the first shell around the counter ion.

The destabilization of the premicellar aggregates at high water content may give rise to a) separation of liquid water, b) formation of inverse micelles, or c) separation of a lamellar liquid crystal. Approximate calculations using the Tanford-Ninham approach gave correct information for a model system, but the critical ratio appeared too insensitive to the alcohol/soap ratio to be useful.

Literature Cited

1. McBain, J. W.; Laing, M. E.; and Titley, A. F. J. Chem. Soc. 115, 1279 (1919).

2. Jones, E. R.; Bury, C. R. Phil. Mag. 4, 1 (1927).

3. Ekwall, P. Acta Acad. Aboensis (Math et Phys.) 4, 1 (1927).

4. Luzzati, V.; Mustacchi, H.; Skoulios, A.; Huson, F. Acta. Crystallogr. 13, 660 (1960).

5. Ekwall, P. "Advances in Liquid Crystals" (G.H. Brown, Ed.) Vol. 1, Academic Press, New York (1974) page 1.

6. Chapman, D. "The Structure of Lipids" Methnen and Co., London (1965).

7. Larsson, K. "Surface and Colloidal Science" (E. Matijevic, Ed.) Vol. 6, John Wiley and Sons (1973) p. 261.

8. Israelachvili, J. N.; Mitchell, D. J.; Ninham, B. W. J. Chem. Soc. Faraday Trans. II, 72, 1525 (1976).

9. Jönsson, B.; Wennerström, H. J. Colloid Interface Sci., 80, 482 (1981).

10. Parsegian, V. A. Trans. Faraday Soc., 62, 848 (1966).

11. Ekwall, P.; Mandell, L.; Fontell, K. Mol. Cryst., 8, 157 (1969).

12. Eicke, H. F.; Christen, H. J. Colloid Interface Sci., 46, 417 (1974).

13. Ruckenstein, E.; Chi, J. C. J. Chem. Soc. Faraday Trans II, 71, 1680 (1975).

14. Ruckenstein, E. J. Dispersion Sci. Technol., 2, 1 (1981).

15. Brois, J.; Bothorel, P.; Clin, B.; Lalanne, P. J. Dispersion Sci. Technol., 2, 67 (1981).

16. Ninham, B. W.; Mitchell, D. J. J. Chem. Soc. Faraday Trans. II, 76, 201 (1980).

17. Christenson, H.; Larsen, D. W.; Friberg, S. E. J. Phys. Chem., 84, 3633 (1980).

18. Friberg, S.; Buraczewska, I. Progr. Colloid Polymer Sci., 63, 1 (1978).

19. Sjöblom, E.; Friberg, S. E. J. Colloid Interface Sci., 67, 16 (1978).

20. Stenius, P.; Rosenholm, J. B.; Hakala, M. R. "Colloid and Interface Science II," (M. Kerker, Ed.) Academic Press, New York, (1976) p. 397.

21. Fendler, J. Acc. Chem. Res., 9, 153 (1976); 13, 7 (1980).

22. Friberg, S. E.; Ahmad, S. I. J. Phys. Chem., 75, 2001 (1971).

23. Friberg, S. E.; Mandell, L.; Ekwall, P. Kolloid-Z.u.Z. Polymere, 233, 955 (1969).

24. Söderlund, G.; Friberg, S. E. Z. Physik. Chem., 70, 39 (1970).

25. Pople, J. A.; Santry, D. P.; Segal, G. A. J. Chem. Phys., 43, 129 (1965).

26. Bendiksen, B.; Friberg, S. E.; Plummer, P. J. Colloid Interface Sci., 72, 495 (1979).

27. Bendiksen, B., Ph.D. Thesis, Chemistry Department, University of Missouri-Rolla, 1981.

28. Schuster, P. "The Hydrogen Bond" (P. Schuster, G. Zundel and C. Sondarfy, Ed's) Vol. I, North Holland, New York, (1976) p. 120.

29. Ekwall, P.; Mandell, L. Kolloid-Z.u.Z. Polymere, 233, 936 (1969).

30. Clausse, M.; Sheppard, R. J.; Boned, C.; Essex, C. G. "Colloid and Interface Sci.", Vol. II (M. Kerker, Ed.) Academic Press, New York (1976) p. 233.

31. Zulauf, M.; Eicke, H. F. J. Phys. Chem., 83, 480 (1979).

32. Jean, Y. C.; Ache, H. F. J. Am. Chem. Soc., 100, 6320 (1978).

33. Jean, Y. C.; Ache, H. F. J. Phys. Chem., 82, 811 (1978).

34. Handel, E. D.; Ache, H. F. J. Chem. Phys., 71, 2083 (1979).

35. Plummer, P. L. M. J. Glaciol., 85, 565 (1978).

36. Santry, D. P.; Crane, R. W. J. Chem. Phys., 2, 304 (1973).

37. Scott, B. F. Theor. Chim. Acta (Berlin) 38, 85 (1975).

38. Reiss, H. J. Colloid Interface Sci., 53, 61 (1975).

39. Tanford, C. "The Hydrophobic Effect", John Wiley and Sons, New York (1973).

RECEIVED August 21, 1981.

2

Radical Pair Reactions in Micellar Solution in the Presence and Absence of Magnetic Fields

NICHOLAS J. TURRO, JOCHEN MATTAY, and GARY F. LEHR

Columbia University, Chemistry Department, New York, NY 10027

The photolyses of dibenzyl ketones in aqueous micellar solution have been shown to greatly enhance both geminate radical pair recombination and the enrichment of ^{13}C in recovered ketone compared to homogeneous solution. These observations have been attributed to the combined effects of the reduced dimensionality imposed by micellization and hyperfine induced intersystem crossing in the geminate radical pairs. This latter effect is the basis of Chemically Induced Dynamic Nuclear Polarization (CIDNP), a phenomenon which is well known in homogeneous solution. The photolyses of 1,2-diphenyl-2-methyl-1-propanone and its ^{2}H and ^{13}C derivatives in micellar solution are now described and further demonstrate the enhanced cage and magnetic isotope effects of micellization. We report also the observation of CIDP during the photolyses of micellar solutions of several ketones, and demonstrate the validity of the radical pair model to these systems. Analyses of the CIDNP spectra in the presence and absence of aqueous free radical scavengers (e.g., Cu^{2+}) allow us to differentiate between radical pairs which react exclusively within the micelle and those that are formed after diffusion into the bulk aqueous phase. In some cases this allows us to estimate a lifetime associated with the exit of free radicals from the micelles.

As evidenced by this symposium, the use of micelles and other organized assemblies to control the selectivity of chemical reactions has recently attracted much attention. In most of these cases, micelles or vesicles have been used as a means of separating charged intermediates formed by electron transfer reactions, thereby preventing the back reaction. The effects of the micelle or vesicle are usually dramatic.

0097-6156/82/0177-0019$05.00/0

The micelle can also be used to enhance the reaction probability of intermediates if they are seqestered inside the micelle. In this case, the micelle acts as a reaction vessel with molecular dimensions. Below we will describe some of our results on the effect of micellization on radical pair reactions. We will show as well that the effects of micellization can be dramatically altered by the application of small external magnetic fields.

The Cage Effect

When 1,2-diphenyl-2-methyl-1-propanone, 1, is irradiated with UV light, the primary photochemical process leads to α-cleavage from the lowest $^3n,\pi^*$ state to give a geminate triplet radical pair.[1,2] In homogeneous solution these fragments begin to separate (10^{-10}s),[3] and lead eventually to products which arise primarily from scavenging or termination reactions of free radicals (Scheme I). Table I lists the chemical and quantum yields for these products in benzene, acetonitrile and methylene chloride. Also listed in Table I are the corresponding yields of products when the photolysis is conducted in cationic, anionic or non-ionic aqueous micellar solution. In these cases, the higher yields of products which could arise from the disproportionation of the geminate radical pair, 2 and 3, as well as the lower quantum yields for disappearance of 1 lead us to suspect that a substantial increase in cage reaction occurs on going from homogeneous to micellar environments.

We have found that addition of Cu(II) to aqueous solutions of HDTCL leads to the selective scavenging of free radicals which have entered the aqueous phase.[2,4] Photolysis of 1 under these conditions leads to the results listed in Table II. These results clearly indicate the enhanced cage effect induced by photolysis of 1 in micellar solution. From the copper scavenging results[4] the cage effect in HDTCL is calculated to be 30% (% cage = moles of 2 or 3/moles of 1 consumed). 2 and 3 appear to be formed exclusively within the micelle by disproportionation of the geminate radical pair. Since cage disproportionation involves reaction of a singlet radical pair, cage reaction cannot occur prior to intersystem crossing (ISC) of the initially geminate triplet pair, the rate of escape of the radicals from the micelle must be slower than the rate of ISC. These results are in good agreement with the observed cage effects of other ketones studied in our laboratory.[4]

The Radical Pair Mechanism

A simple model which accounts for the observed cage effects[5] in micellar solution is diagrammed in Figure 1. Fragmentation of a triplet molecular species (produced by light absorption to form a singlet followed by ISC to a triplet) proceeds along the repulsive energy potential until the fragments are

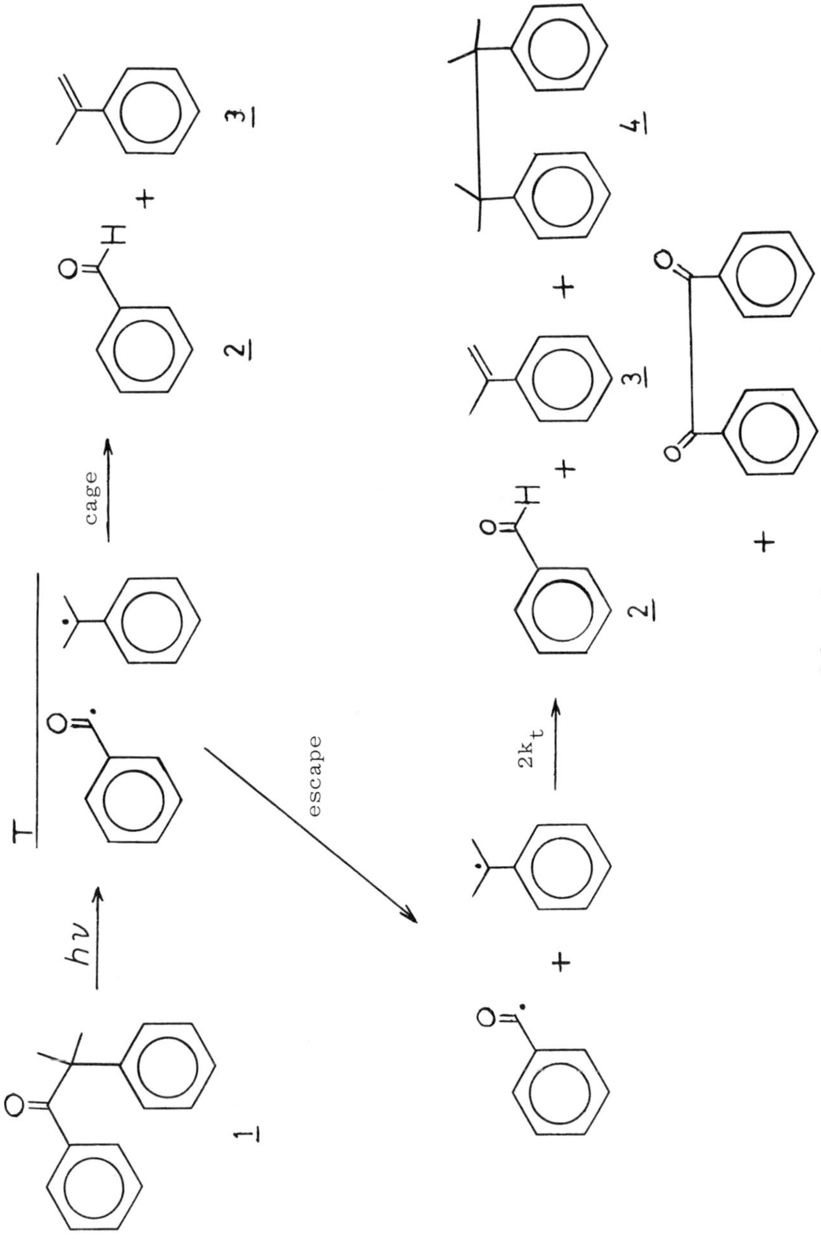

Scheme I

Table I. Relative Yields (based on ketone consumed) and Quantum Yields for the Photo Products of 1 in Homogeneous and Micellar Solutions.

Solvent or Detergent	ϕ-ket	2		3		4		5	
		%	ϕ	%	ϕ	%	ϕ	%	ϕ
C_6H_6	0.87	15	0.13	3.4	0.03	21	0.18	8	0.07
CH_3CN	0.95	10	0.10	3.3	0.03	18	0.17	9	0.09
CH_2Cl_2	0.97	18	0.18	6	0.06	18	0.18	7	0.07
HDTCl[(a)]	0.73	22	0.16	23	0.16	2.2	0.02	-	--
HDTBr[(b)]	0.73	27	0.19	23	0.16	3.5	0.03	-	--
DDTCl[(c)]	0.75	24	0.17	15	0.11	4.0	0.03	-	--
SDS[(d)]	0.67	30	0.19	17	0.11	2.8	0.02	-	--
Brij 35	0.88	22	0.19	17	0.15	2.6	0.02	-	--

(a) Hexadecyltrimethylammonium chloride

(b) Hexadecyltrimethylammonium bromide

(c) Dodecyltrimethylammonium chloride

(d) Sodium dodecyl sulfate

Table II. Relative Yields (based on ketone consumed) and Quantum Yields for the Photo Products of 1 in HDTCl Micellar Solution in the Presence of $CuCl_2$.

Cu(II)/1	ϕ_{-ket}	2 %	2 ϕ	3 %	3 ϕ	4 %	4 ϕ
0	∿0.73	22	0.16	23	0.16	2.2	0.02
0.4	∿0.7	(a)	0.19	(a)	0.21	(a)	0.015
2	∿0.7	(a)	0.20	(a)	0.19	(a)	0.01
4	∿0.7	(a)	0.20	(a)	0.20	(a)	trace
5[b]	∿0.7	29	0.20	30	0.21	0.0	0.00

(a) Not measured.

(b) Yields of scavenging products: benzoic acid, $\geq$33% ($\phi \geq 0.23$); α,α dimethylbenzyl chloride, 26% ($\phi = 0.18$); α,α dimethylmethyl benzyl alcohol, 23% ($\phi = 0.16$).

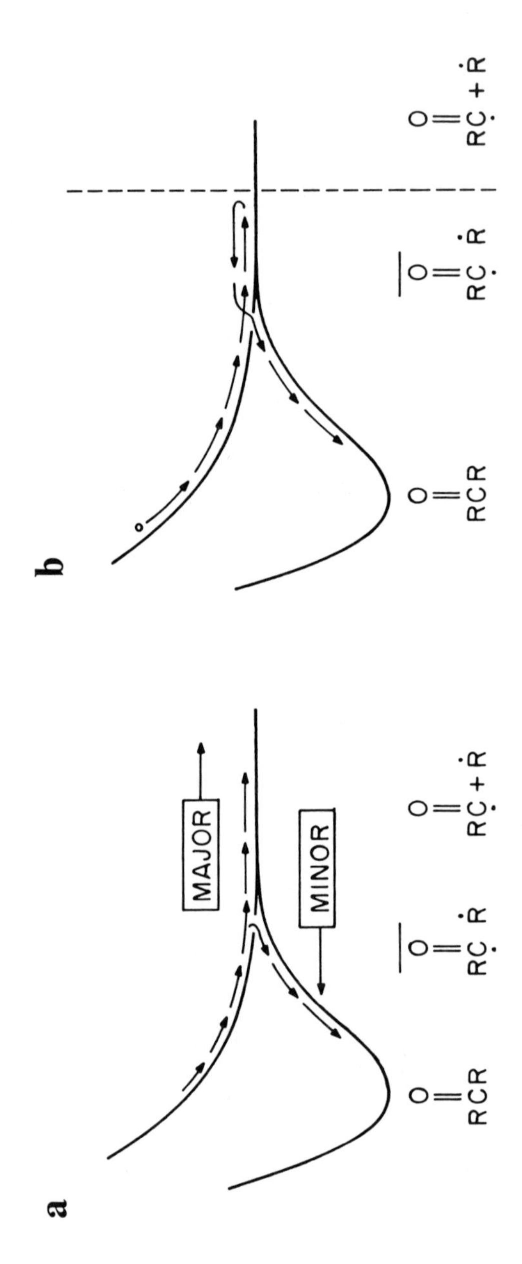

Figure 1. Potential energy surfaces for singlet and triplet radical pairs: a, without and b, with a reflecting wall.

separated to a distance where the exchange interaction of the unpaired electrons is negligible. At this point, the singlet (S) and triplet (T) energies are essentially degenerate and ISC within the radical pair can occur. In homogeneous solution the random movement of the radical fragments most often leads to irreversible separation of the pair,[3] and we therefore expect most of the products to arise from uncorrelated free radicals. The enhanced cage effect in micellar solutions arises because the chemical potential of the micelle-water interface provides a reflecting wall in the region of the S and T degeneracy. The radical fragments are prevented from diffusing beyond several angstroms from their original location, and the chances of a reencounter are therefore increased. The radical pair thus experiences a longer period of time to effect ISC from T to S.

The mechanism by which S and T can interconvert is based on the radical pair model of Chemically Induced Dynamic Nuclear Polarization (CIDNP).[6] According to this theory, nuclear magnetic moments which are coupled to the unpaired electrons by hyperfine interactions (A_{hfs}) can alter the rate at which the electron spins lose their phase coherence, i.e., undergo ISC. In the presence of an applied magnetic field greater than the hyperfine splittings, the components of the triplet surface split into T_+, T_0 and T_- levels, in which the degeneracy of T_+ and S is removed (Figure 2). The effect of a strong applied field then is to inhibit ISC of ∿2/3 of the triplet radical pairs from the ensemble which could (at zero field) ISC to S. Several predictions based on the radical pair model[7] can be made with regard to the probability of cage reaction of a radical pair:

(a) for a triplet radical pair, the cage effect at high field ($H_0 > A_{hfs}$) will be less than that observed at low or zero field ($H_0 \lesssim A_{hfs}$).
(b) intersystem crossing will be fastest for radical pairs with the largest magnitude of hyperfine coupled nuclear spins.
(c) a larger hyperfine coupling will result in faster intersystem crossing.
(d) the maximum intersystem crossing rate occurs at an applied field approximately equal to the hyperfine splittings.

It was recognized[7a] very early in the development of the radical pair model that a natural consequence of (b) was that the separation of nuclear spin isotopes from non-magnetic isotopes was possible. Since the cage reaction in homogeneous solution contributes very little towards the overall product yields, the "magnetic isotope effect" is not very significant in homogeneous solution.[4,8] In micellar solution, the extent of cage reaction can be substantially increased. Below we will examine the consequences of an applied magnetic field on the photoreactions of ketones in micellar solution.

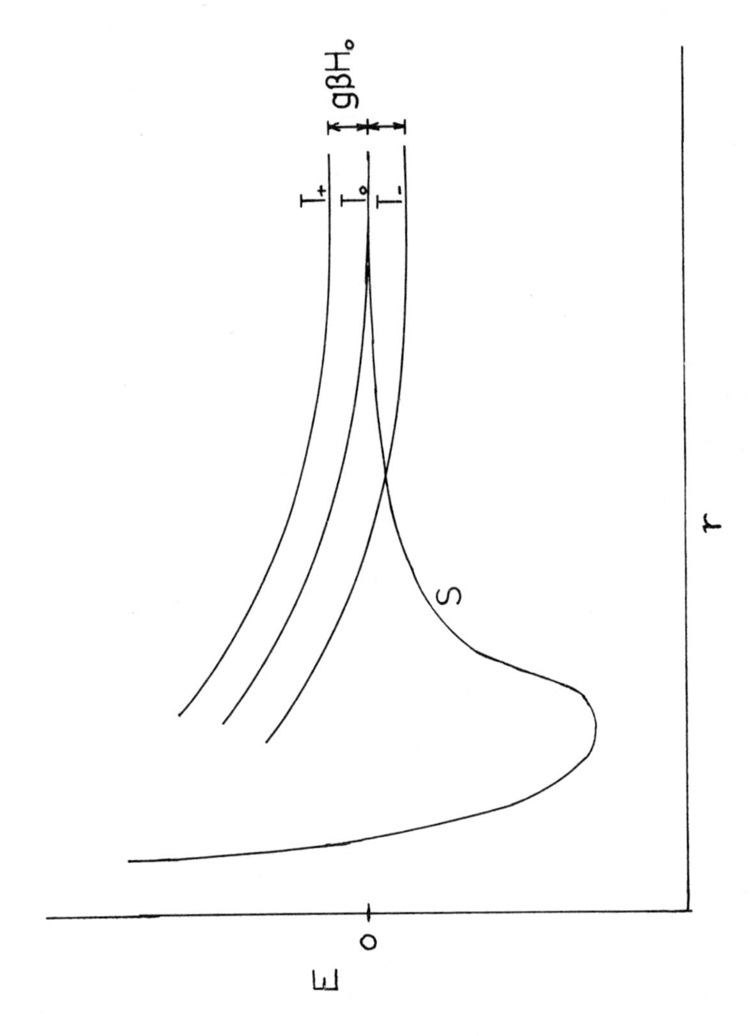

Figure 2. Potential energy surface for a radical pair in a strong magnetic field showing the splitting of the T_+, T_0, and T_- levels.

Magnetic Field Effects on Cage Efficiencies

Figure 3 shows the effect of applied magnetic field on the cage effect for 1 and some of its isotopic isomers. The higher cage effect for ^{13}C enriched ketone (compound 8 in Figure 3) correlated well with prediction (b). In this case, we have added an additional spin 1/2 nucleus with a large hfs ($A_{^{13}C}$=128 Gauss)[9] to one of the radical fragments. Likewise, the lower cage effect observed for the deuterated ketones (6 and 7 in Figure 3) correlates well with prediction (c) and the relative hfs of ^{1}H and ^{2}H (1:4). Its noteworthy that 2 shows no mass isotope effect for the disproportionation reaction.[5]

With the application of a magnetic field, the cage effect for 1 drops substantially. The full effect (30% decrease) is achieved with the application of a field of only a few hundred Gauss, the magnitude of the field commonly available from a magnetic stirring bar. A similar decrease is observed for 6 and 7. The magnetic field effect on 8 is notably different from 1. In this case the cage effect remains unchanged over the first 100 Gauss before the fall off occurs. Taking into account prediction (d), we would expect the larger hfs of the carbonyl ^{13}C of the benzoyl radical ($A_{^{13}C}$=128 Gauss) to continue to provide an effective ISC pathway up to approximately 100 G. The increased ISC rate due to ^{13}C, however, is counterbalanced by a decreasing ISC rate to ^{1}H as the field becomes significantly larger than the proton hfs ($A_{^{1}H}$=16G). At an applied field of 1000 G or greater, the cage effect reaches a steady plateau and shows no further magnetic field effect.

The Magnetic Isotope Effect

In order to efficiently enrich a cage product in a magnetic isotope such as ^{13}C, several conditions should be satisfied.[7] Most effective enrichment will result when (1) the radical pair is produced exclusively in the triplet state, (2) there is significant probability of cage reaction, i.e., a substantial cage effect, (3) there is only one product of the cage reaction, i.e., recombination or disproportionation, (4) the cage and escape products are clearly different and easily separable. These criteria have been met by several compounds.[8,10] One of the most widely used substances is dibenzyl ketone (DBK). Upon direct photolysis, a triplet radical pair consisting of a phenylacetyl and a benzyl radical is formed via α-cleavage of the excited triplet ketone.[11] The principal cage reaction results in recombination to the starting ketone, with a small amount of "head-to-tail" reaction to give p-methylphenyl acetophenone (PMPA). Phenyl acetyl radicals which do not recombine by rapid ISC and recombination suffer unimolecular decomposition to benzyl radicals and CO with a half life of about 10-100 ns.[11] Thus, even in the absence of free radical scavengers, random termination of the escaped radicals leads to a unique product, diphenylethane (DPE).

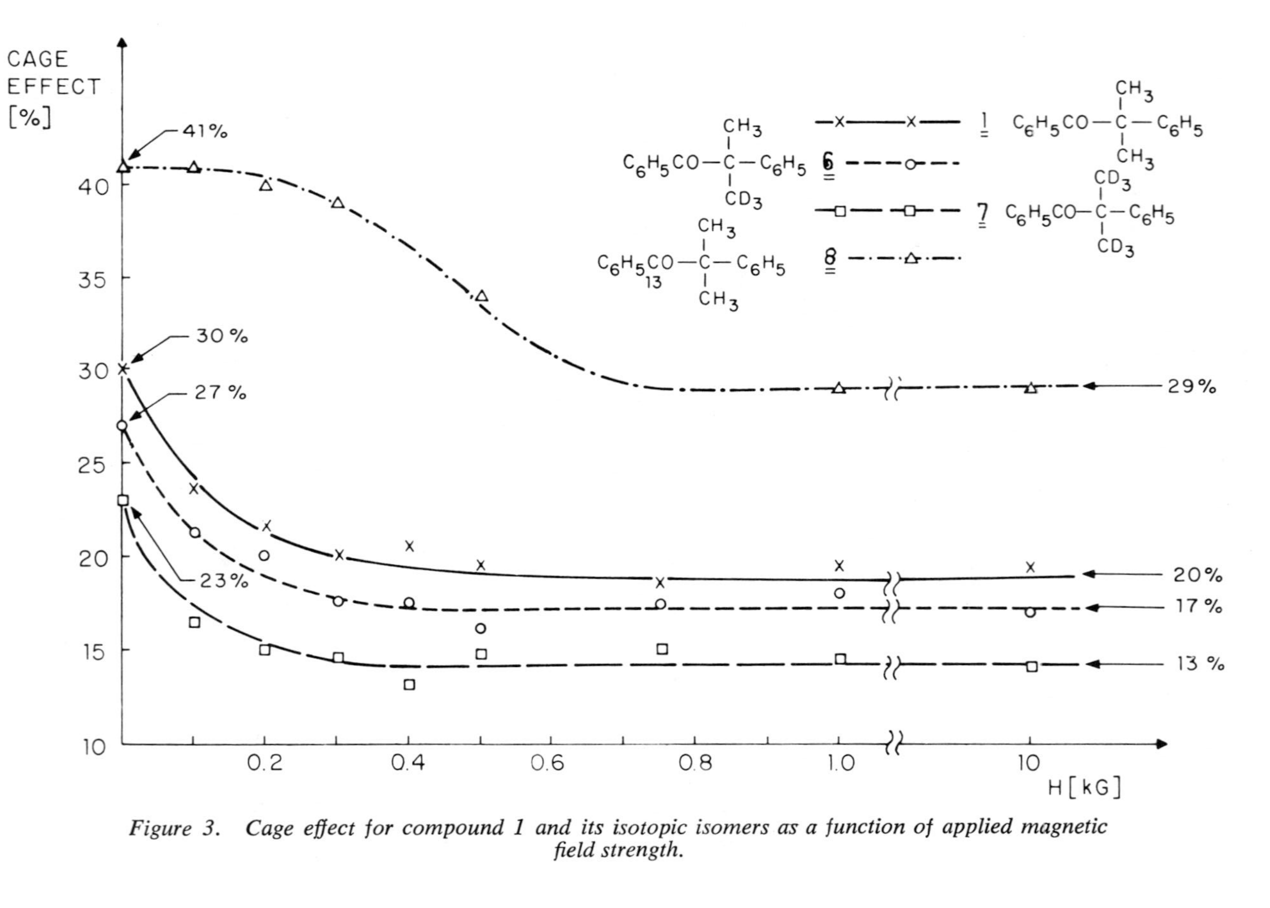

Figure 3. Cage effect for compound 1 and its isotopic isomers as a function of applied magnetic field strength.

When DBK is photolyzed in homogeneous solution, a small isotopic enrichment of ^{13}C in the recovered ketone is observed.[(8b,12)] The extent of enrichment is dependent on the solvent viscosity. When DBK is photolyzed in HDTCl micellar solution, however, the efficiency of ^{13}C enrichment in the recovered ketone increases dramatically. Measurement of the enrichment efficiency can be expressed in terms of Bernstein's parameter, α. Evaluation of

$$\alpha = \frac{\text{probability of product formation for } ^{13}C}{\text{probability of product formation for } ^{12}C}$$

this parameter by either mass spectrometric analysis or by determination of the ratio of quantum yields of reaction of ^{12}C and ^{13}C containing DBK[13] (^{13}C containing DBK was 90% enriched at the carbonyl carbon)[(4)] results in the data presented in Figure 4. At zero field, enrichment of ^{13}C in the starting ketone is substantial, and correlates well with some of the micelle properties discussed above and the radical pair model. Most importantly, the enrichment as a function of applied field goes through a maximum between 150 and 300 G as predicted by the radical pair model. Deuteration of the methylene groups does not alter the enrichment at zero field, and, interestingly, does not lead to a maximum in the field dependence.

CIDNP in Micellar Solution

If we are to use the radical pair theory to explain the effects of micellization on the cage reaction probability as well as the magnetic field effect, it is mandatory that we be able to observe CIDNP in these systems. In addition, since CIDNP is sensitive to events on the time scale of the radical pair lifetime, detailed analysis of the CIDNP can often lead to mechanistic insight to the dynamics of the radical pair. Below we describe one such result.[(11c)]

As shown above, the process of diffusion in the micelle environment is restricted by the chemical potential of the micelle water interface. On the short time scale (0-1000ns), the radical fragments are kept in close proximity to one another, so that the probability of a reencounter is high. Micelles are not, however, static systems. In addition to exchanging monomer surfactant molecules, solubilized molecules or intermediates are subject to exchange. For most small or moderate sized hydrophobic fragments, the rate of escape from the micelle is given by first order rate constants in the range 10^5-10^7 sec^{-1}, while the rate of re-entry is generally diffusion controlled.[(16)] Thus, the rate of exchange of fragments among micelles is given by the exit rate of the probe.

When di-t-butyl ketone is photolyzed in homogeneous solution[(14)], the CIDNP spectrum shown in Figure 5a is observed. Interpretation of this spectrum is based on Scheme II. α-Cleavage

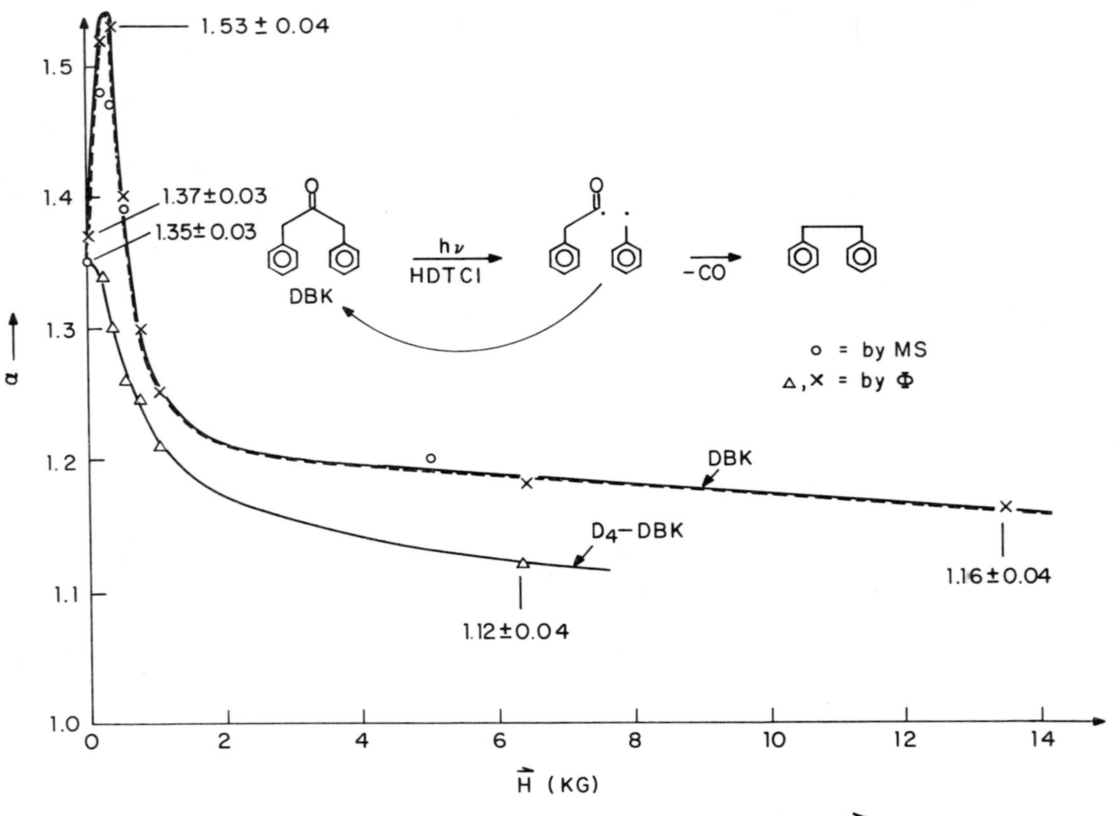

Figure 4. ^{13}*C-enrichment parameter at a given magnetic field* $\vec{H}$.

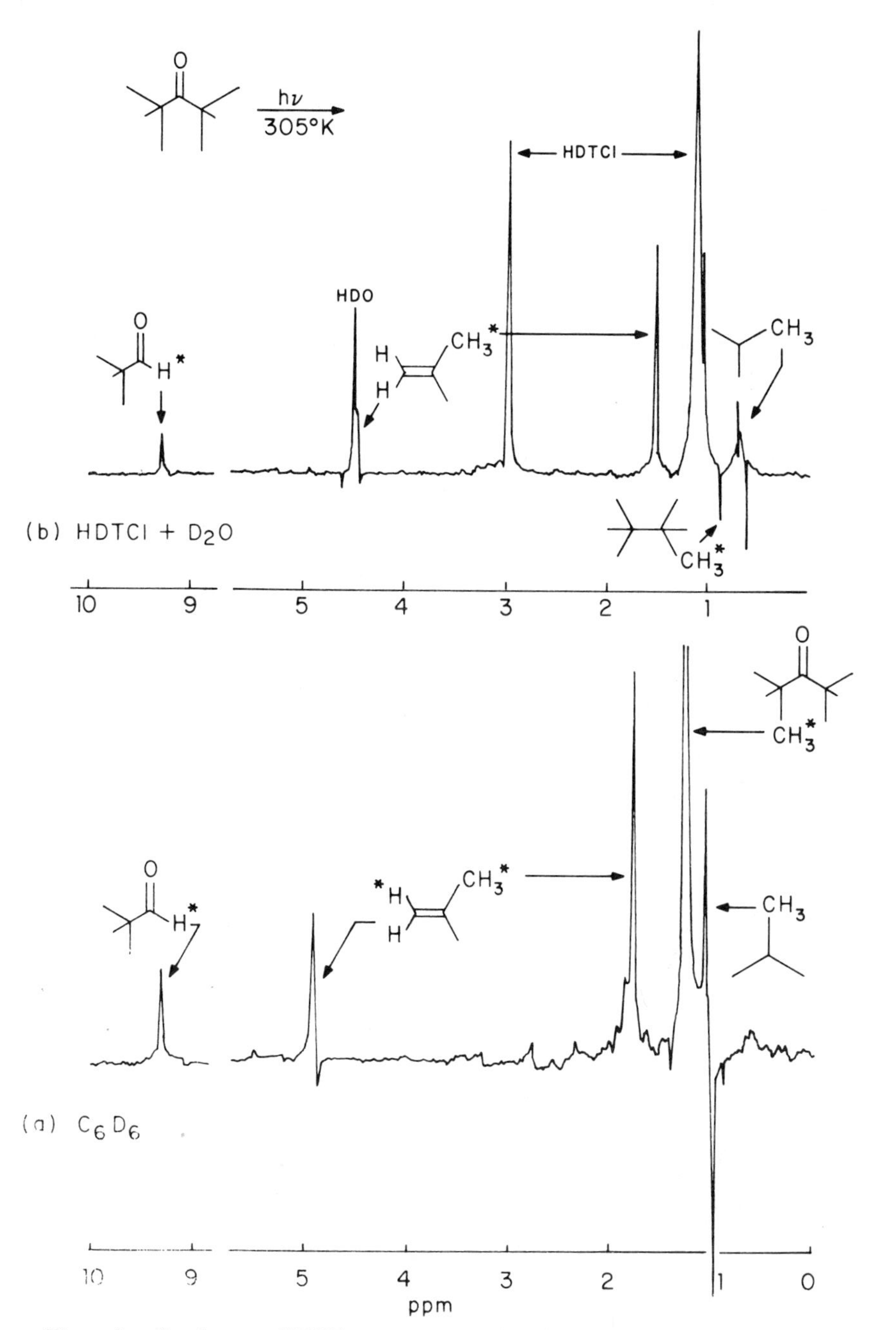

Figure 5. Steady state CIDNP spectra during the photolysis of di-t-butyl ketone in a, homogeneous C_6D_6 solution and b, HDTCl micellar solution.

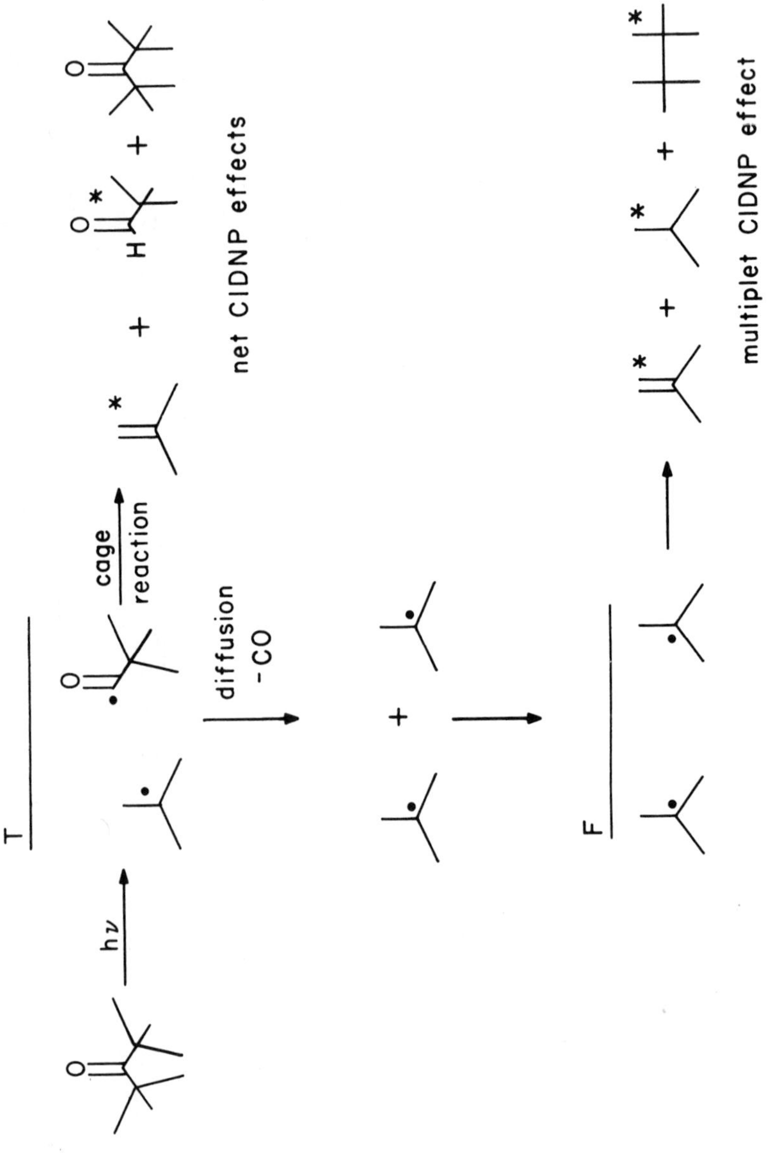

hν

T

cage reaction

net CIDNP effects

diffusion -CO

F

multiplet CIDNP effect

Scheme II

from the excited triplet state results in a triplet radical pair (T-pair) consisting of pivaloyl and t-butyl. The cage products of this pair, pivaldehyde, isobutylene and the starting ketone, will exhibit net CIDNP effects, since the radicals have different g-factors[(6)] (i.e., $\Delta g \neq 0$). Most of the radicals diffuse into the bulk solution to become truly free radicals. The lifetime of the pivaloyl radical at room temperature is only a few microseconds.[(15)] Thus, decarbonylation occurs prior to the random encounter of free radicals to form free radical pairs (F-pairs). The F-pairs, consisting of two t-butyl radicals, may also result in products which exhibit CIDNP effects. Since there is no g-factor difference we expect to observe multiplet effects in these products.[(6)] These expectations are confirmed[(14)] in the spectrum shown in Figure 5a. Likewise, when this photolysis is conducted in HDTCL micellar solutions in D_2O[(11c)], an essentially identical CIDNP spectrum is observed, as shown in Figure 5b.

The rate of decarbonylation of pivaloyl[(15)] is within the range of rates we expect for the exchange of radicals between micelles.[(16)] The question arises as to whether or not exchange takes place prior to or after decarbonylation of the pivaloyl radical. This is shown in Scheme III.

By addition an aqueous phase radical trap, we can intercept any radicals which enter the water phase and prevent the formation of F-pairs. Figure 6 shows the effect of added Cu^{2+} on the high field region of the CIDNP spectrum. The disappearance of the multiplet CIDNP in isobutane and the residual net polarization in 2,2,3,3-tetramethylbutane is complete with the addition of only 0.5 equivalents of Cu^{2+}. The net CIDNP in isobutylene, however, is not affected by the addition of Cu^{2+}. Thus, we are able to scavenge t-butyl radicals before they can form F-pairs, but do not affect the T-pairs of pivaloyl and t-butyl radicals.

If we estimate the time required for S-T mixing and recombination for a radical pair to be 100 ns[(6)] and the lifetime of the pivaloyl radical at $31^{o}C$ to be ~ 6 μs[(15)], we can estimate the rate constant for the exit of t-butyl/pivaloyl radicals from HDTCl micelles to be on the order of 10^6-10^7 sec^{-1}. This is nicely in line with exit rates of small phosphorescent probe molecules from similar micelle systems.[(16)]

Acknowledgements

The authors wish to thank the National Science Foundation and the Air Force Office of Scientific Research for their generous support of this work. J. M. thanks the Deutscher Akademischer Austauschdienst for a NATO Fellowship. M. Ressel-Mattay is warmly acknowledged for her assistance in the synthesis of 1, 6, 7 and 8.

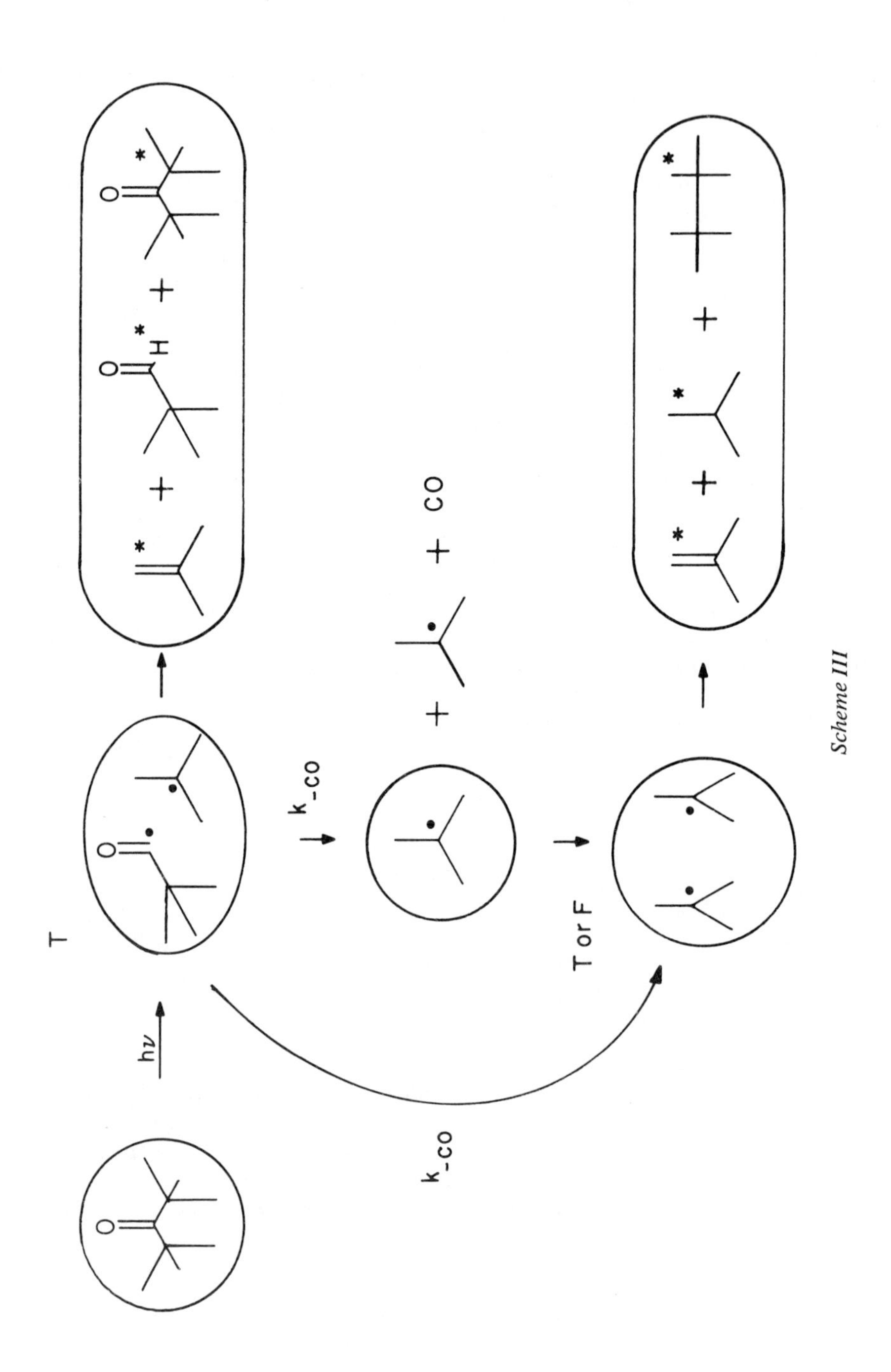

Scheme III

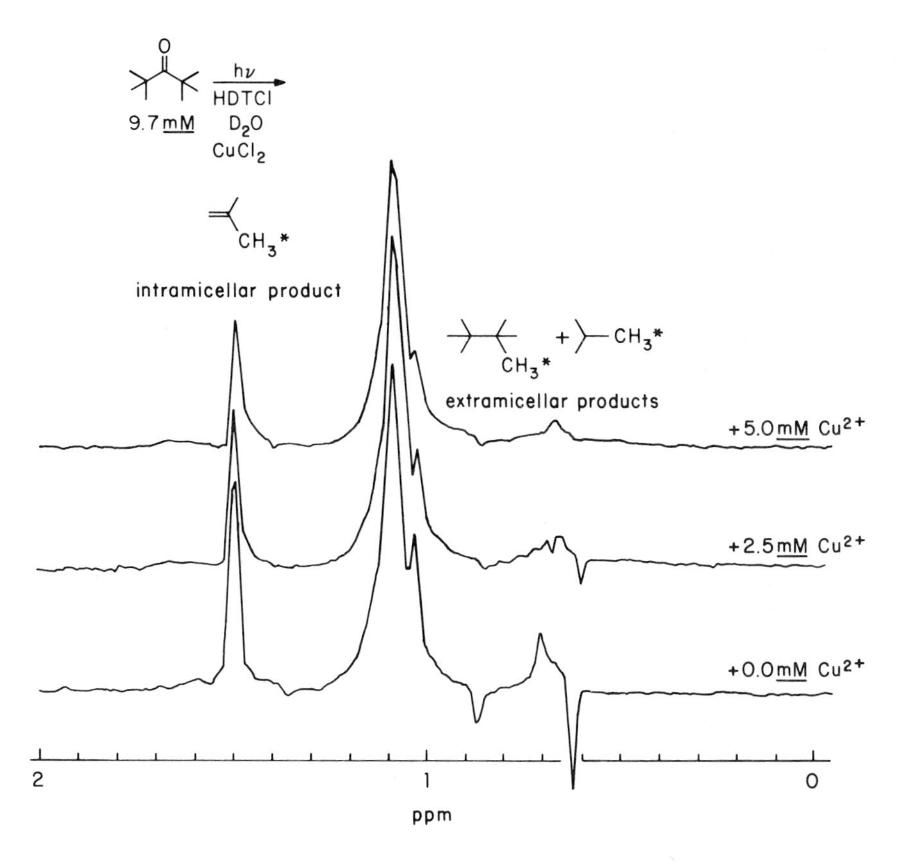

Figure 6. Steady state CIDNP spectra during the photolysis of di-t-butyl ketone in the high-field region in HDTCl micellar solution in the presence of $CuCl_2$.

Literature Cited

1. Heine, H.-G.; Hartman, W.; Kory, D.R.; Magyar, J.G.; Hoyle, C.E.; McVey, J.K.; and Lewis, F.D. J. Org. Chem. 1974, 39, 691.
2. Turro, N.J. and Mattay, J. Tet. Lett. 1980, 1799.
3. Noyes, R.M. J. Am. Chem. Soc. 1956, 78, 5486.
4. Turro, N.J.; Chow, M.-F.; Chung, C.-J.; Weed, G.C. J. Am. Chem. Soc. 1980, 102, 4383.
5. Turro, N.J. and Mattay, J. J. Am. Chem. Soc., in press.
6. (a) Lawler, R.G. Prog. NMR Spectr. 1973, 9, 145.
   (b) Kaptein, R. Adv. Free Rad. Chem. 1975, 5, 381.
7. (a) Lawler, R.G. and Evans, G.T. Ind. Chim. Belge, 1971, 36 1087.
   (b) Buchachenko, A.L. Russ. Chem. Rev. 1976, 45, 761.
8. (a) Buchachenko, A.L. Russ. J. Phys. Chem. 1977, 51, 2461.
   (b) Sterna, L.; Ronis, D.; Wolfe, S. and Pines, A. J. Chem. Phys. 1980, 73, 5493.
9. Landolt-Bernstein, "Organic C-Centered Radicals", Berndt, A.; Fischer, H. and Paul, H. Eds., Vol. 9, Part b, Springer-Verlag, New York 1977.
10. Turro, N.J.; Kraeutler, B.; Acc. Chem. Res. 1980, 13, 369.
11. (a) Brunton, G.; McBay, H.C.; and Ingold, K.U. J. Am. Chem. Soc. 1977, 99, 4447.
    (b) Robbins, W.K. and Eastman, R.H. J. Am. Chem. Soc. 1970, 92, 6077.
    (c) Lehr, G.F. and Turro, N.J. Tetrahedron, in press.
12. Turro, N.J. and Kraeutler, B. J. Am. Chem. Soc. 1978, 100 7432.
13. Kraeutler, B. and Turro, N.J. Chem. Phys. Lett. 1980, 70 266.
14. Tomkiewicz, M.; Groen, A. and Cocivera, J. J. Chem. Phys. 1972, 56, 5850.
15. Schuh, J.; Hamilton, E.J.; Paul, H. and Fischer, H. Helv. Chim. Acta, 1974, 57, 2011.
16. Almgren, M.; Grieser, F. and Thomas, J.K. J. Am. Chem. Soc. 1979, 101, 279.

RECEIVED July 2, 1981.

3

Light-Induced Electron Transfer Reactions of Metalloporphyrins and Polypyridyl Ruthenium Complexes in Organized Assemblies

D. G. WHITTEN, R. H. SCHMEHL, T. K. FOREMAN,
J. BONILHA, and W. M. SOBOL

The University of North Carolina at Chapel Hill, Department of Chemistry,
Chapel Hill, NC 27514

We have prepared and studied a number of surfactant, hydrophobic and water soluble luminescent metal complexes. These can serve as excited substrates in light-induced electron transfer reactions. Both the quenching processes and subsequent reactions can be strongly affected by incorporation of the substrate and/or quencher in an organized assembly. This paper focuses mainly on studies in micelles. In our investigations, we have attempted to control the light-induced redox reactions through the use of the organized assemblies so that net useful chemical conversions can be obtained. We have also used some of these reactions as a probe to study the structure and binding properties of surfactant micelles, particularly the micelle-water interface region.

Light-induced electron transfer reactions, especially those occurring through visible light excitation, have been the focus of extensive investigation over the past several years (1-21). Recently there has been a great deal of study involving the use of organized assemblies, particularly aqueous surfactant media such as micelles or bilayer vesicles to modify the course of these reactions (22-34). While many interesting results have been obtained in several of these studies, it is clear that much remains to be done to optimize conditions for obtaining efficient charge separation or practical net chemical conversion. The present paper will focus on some recent studies carried out in our

0097-6156/82/0177-0037$05.00/0

laboratories using luminescent metal complexes as excited substrates and a variety of electron acceptors as quenchers. The basic photochemical process and energy wasting back reactions (eqs. 1-3) have been elaborated in numerous studies for most of these systems in homogeneous solution (1-21). The

$$MC \xrightarrow{h\nu} MC^* \qquad (1)$$

$$MC^* + A \longrightarrow MC_{ox} + A_{red} \qquad (2)$$

$$MC_{ox} + A_{red} \longrightarrow MC + A \qquad (3)$$

current work describes the modification of rates and efficiencies of these and competing reactions when either the substrate, quencher or both are associated with an organized assembly such as a surfactant micelle. The results show that through systematic selection of surfactant, substrate and quencher the light-induced electron transfer process can be diverted to drive efficient chemical reactions in which both the substrate and quencher can be recycled. Moreover, our studies show that these reactions can also be used to investigate the structure of the organized assemblies themselves, particularly with respect to their interfacial properties.

Intramicellar Electron Transfer Quenching

Some of our initial studies in this area have involved the use of the surfactant ruthenium complexes 1 and 2 and the electron acceptor N,N'-dimethyl-4,4'-bipyridinium (methylviologen) (MV^{2+}) (28). Previously we and others (4,5) had

R R 2+
N N
Ru
(bipy)$_2$

1 $R = C_{18}H_{37}$

2 $R = C_{16}H_{33}$

demonstrated that quenching of tris (2,2'-bipyridine) ruthenium $(II)^{2+}$, ($Ru(bipy)_3^{2+}$), excited states by MV^{2+} occurs as outlined in eqs. 1-3. Solubilization of 1 or 2 in anionic (sodium dodecylsulfate (SDS)) or cationic (cetyltrimethyl ammonium bromide (CTAB)) micellar solution occurs readily although the complexes are totally insoluble in water in the absence of detergent. Consequently the surfactant complexes must be totally associated with the micellar phase. Addition of MV^{2+} to the aqueous micellar solutions of 1 or 2 results in strong quenching in the case of SDS micelles but very little quenching for the CTAB micelles (Table I).

As Table I indicates, Stern-Volmer quenching constants obtained in SDS by measuring either excited state lifetimes or luminescence intensities were within experimental error, moreover the observed quenching constants are strongly dependent on surfactant concentration. The decay of the

Table I. Quenching of Surfactant Ruthenium Complex Luminescence by MV^{2+} in Various Media

Complex	Medium	μ^a	Observed Quenching Constant, k_{SV}	
			$I°/I$	$\tau°/\tau$
1,2	CH_3CN	0.20	260	
1	0.00125M CTAB	0.20	<1	
2	0.012M SDS	0.20	4260	4109
2	0.024M SDS	0.20	2403	2250
2	0.042M SDS	0.20	1170	1140

[a] NaCl used as supporting electrolyte.

excited state in SDS containing MV^{2+} was found to follow simple monoexponential decay in each case (29,35). We have interpreted the reduced quenching in CTAB micelles as being due to exclusion of the cationic viologen from the micellar phase due to coulombic repulsion. The enhanced (compared to acetonitrile solution) quenching of 1 and 2 by MV^{2+} in aqueous SDS solutions can be explained in terms of an intramicellar quenching process in which the micelle-solubilized substrate is quenched by MV^{2+} also associated with the anionic surfactant (29). The fact that only monoexponential decay is observed in the system 2, MV^{2+}, SDS, and the dependence of the quenching constant on surfactant concentration suggest a mechanism whereby MV^{2+} is strongly bound to the anionic micelles but exchanging very rapidly between aqueous and micellar pseudophases (26,29). Unlike the situation with $Ru(bpy)_3^{2+}$ - MV^{2+} in water or acetonitrile, where luminescence quenching is accompanied by the generation of moderately long-lived transient ions (4,5), the present intramicellar quenching process does not result in production of detectable amounts of free ions. Presumably, coulombic factors and the micelle "cage effect" inhibit separation of the product ions before back reaction occurs.

In extensions of our studies of reaction of photoexcited 2 with MV^{2+} in the prescence of anionic surfactants we have investigated the effect of added electrolytes on both the binding of MV^{2+} to the assemblies and the intramicellar quenching process (36). Table II gives values for the exchange equilibrium constant k_{ex} (eq. 4) and the intramicellar quenching constant k_q obtained for the system 2 - MV^{2+} in aqueous SDS. The variation of k_{ex} with

$$\bigcirc\begin{matrix} S^+ \\ S^+ \end{matrix} + MV^{2+}_{aq} \underset{}{\overset{K_{ex}}{\rightleftarrows}} \bigcirc MV^{2+} + 2S^+_{aq} \qquad (4)$$

Table II Equilibrium Constants, k_{ex}, and Intramicellar Quenching Constants, k_q, for the System 2, MV^{2+} in Aqueous SDS with Various Cations, S^+

S	K_{ex}	$K_q \times 10^{-7} M^{-1} sec^{-1}$
$(C_2H_5)_4N^+$	279	5.67
Na^+	868	1.93
Mg^{2+}	102	1.36
Li^+	3566	1.28
NH_4^+	5625	0.97

added cations might be attributed to several factors. With the exception of Et_4N^+, the values of k_{ex} increase with decreasing charge/size ratios of the hydrated cations. This could be interpreted as indicating that highly charged small ions bind more readily due to electrostatic attraction and thus compete more effectively with the viologen for binding sites on the micelle. An alternative interpretation could be that the added ions "tighten up" the micelle by reducing head group repulsive interactions. The low value obtained for binding of MV^{2+} in the prescence of the organic cation Et_4N^+ is noteworthy and suggests the role of hydrophobic interactions (vide infra) in interactions of both organic cations with the micelle.

The variation of k_q observed with the several added cations is not very pronounced, with the exception of Et_4N^+. It is noteworthy that, while Et_4N^+ suppresses the binding of MV^{2+} to the micelles, it actually enhances the intramicellar quenching constant by a factor of 3-5 over that obtained with the inorganic cations. A possible explanation for this phenomenon which will be discussed later in more detail (vide infra) is that the Et_4N^+ binds primarily in hydrophobic-hydrophilic sites in the micelle interior such that the viologen is more likely to be excluded from these sites and thus near the surface in its presence, hence "closer" to the hydrophilic substrate chromophore of 2.

Interfacial Electron Transfer Processes in Micellar Solutions

As indicated above, while concentrating both the light absorbing substrate and electron acceptor-quencher in an anionic micelle can lead to an enhanced quenching process, it usually gives an enhanced back electron transfer process as well. As several studies have recently demonstrated, however, the use of surfactant assemblies can result in prolonged charge separation when the assembly binds only one of the substrate-quencher pair, or perhaps more importantly, only one of the products (29,34). One such system we have investigated consists of the water soluble tetra(p-sulfonatophenyl) porphyrins 3 and 4 as excited substrates and various viologen derivatives as electron acceptor-quenchers in the prescence of either cationic or anionic micelles. In homogeneous aqueous solution

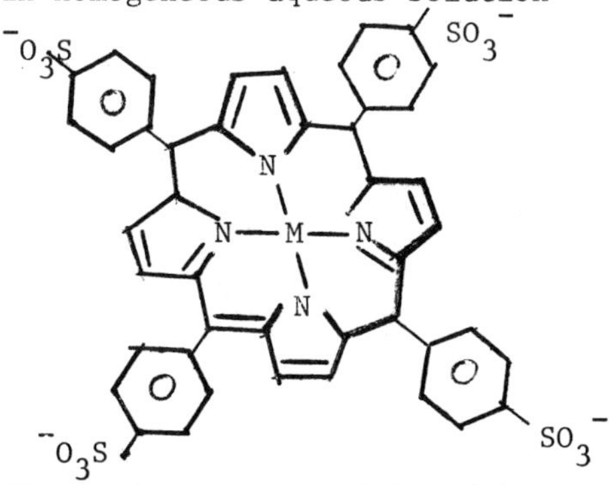

3 M = 2H

4 M = Pd

The excited states of 3 and 4 are efficiently quenched by viologen derivatives such as MV^{2+} or dibenzylviologen (BV^{2+}); both dynamic quenching and static quenching, due to ground state association between the porphyrin and viologen, are observed. No observable charge separation occurs as a result of the quenching since even the product ions are of opposite charge and should strongly associate. Addition of either cationic (CTAC) or anionic (SDS) detergent above the CMC results in attenuated quenching but permits the observation of relatively long-lived product ions (Table III); both the lifetime of the product ions and the rate of the quenching process are strongly affected by the concentration and nature of added electrolytes. It is difficult to precisely and quantitatively define the overall effects produced by adding detergent and salts to the aqueous solutions containing the tetraanionic porphyrins 3 and 4 and the various viologens; however a careful consideration of what can be established with certainty leads to a fairly clear and systematic picture of the overall events occurring as follows:

(1) The tetraanionic porphyrins associate strongly with moderately hydrophobic cations such as CTAC, R_4N^+ and the viologens. These complexes are readily detected spectroscopically for both 3 and 4; however, efforts to measure precise values of the association constant are frustrated due to the fact that multiple association complexes are formed (eqs. 5, 6), where P^{4-} = 3 or 4 and R_4N^+ is either a cationic detergent or simple tetraalkylammonium ion.

$$P^{4-} + R_4N^+ \rightleftharpoons (P\cdot\cdot R_4N)^{3-} \rightleftharpoons [P\cdot\cdot(R_4N)_2]^{2-} \text{ etc.} \quad (5)$$

$$P^{4-} + RV^{2+} \rightleftharpoons (P\cdot\cdot RV)^{2-} \rightleftharpoons [P\cdot\cdot(RV)_2] \quad (6)$$

Table III. Back Reaction Rate Constants and $t_{1/2}$ Values (Flash Photolysis Conditions) for Porphyrin 4 and Viologen Quenchers in Aqueous Detergent Solutions

Surfactant	Salt	Quencher	$k_q \times 10^{-9}$ $M^{-1}s^{-1}$	$k_b \times 10^{-9}$ $M^{-1}s^{-1}$	$t_{1/2}$
none	0.05M NaCl	MV^{2+}	3.7	a	a
SDS	0.05M NaCl	MV^{2+}	0.03	a	a
SDS	0.02M $(CH_3)_4NCl$	MV^{2+}	0.12	2.0	9.2
SDS	0.08M $(C_2H_5)_4NCl$	MV^{2+}	0.27	13.0	1.5
SDS	0.05M $(nC_4H_9)_4NBr$	MV^{2+}	--	44.0	0.2
SDS	0.05M NaCl	BV^{2+}	0.31	2.9	3.9
SDS	0.02M $(C_2H_5)_4NCl$	BV^{2+}	--	0.52	7.3
CTAC	0.05M NaCl	MV^{2+}	0.018	29.5	0.15
CTAC	0.05M $(CH_3)_4NCl$	MV^{2+}	--	10.3	0.57
CTAC	0.05M $(C_2H_5)_4NCl$	MV^{2+}	--	6.3	1.5
CTAC	0.05M NaCl	BV^{2+}	0.31	3.7	5.1
CTAC	0.05M $(CH_3)_4NCl$	BV^{2+}	0.26	1.8	10.0
CTAC	0.05M $(C_2H_5)_4NCl$	BV^{2+}	--	0.2	100.0

[a] not observable

(2) Both the viologen dications and the cations of the added electrolyte associate with anionic (SDS) detergent above the CMC. The more hydrophobic electrolytes (e.g., $(n\text{-}C_4H_9)_4N^+$, $(C_2H_5)_4N^+$) are more effective in reducing the binding of viologens to the micelle (see above as well as later discussion) (36). The reduced viologens RV^+ should also bind strongly to the anionic micelles, and the effect of added hydrophobic ions (R_4N^+) should be more or less parallel to that observed for the viologen dications.

(3) The free porphyrin tetraanions do not associate with anionic micelles due to coulombic repulsions; however, the cation-associated porphyrins (i.e., $(P\cdot R_4N)^{3-}$) should experience less repulsion such that associated forms may either bind or at least encounter the micelle and interact with micelle-associated viologens.

(4) No transients are observed for the porphyrin viologen system in the absence of surfactant regardless of the electrolyte added.

Given the above observations and the data in Table III it becomes clear that in the prescence of micellar SDS the quenching should consist of two processes, interaction of free (or associated) porphyrin excited states with viologen in the aqueous phase (eq. 7) and reaction of cation-associated porphyrin excited states with micelle-bound viologen (eq. 8). Reaction 7 will give

$$P^*(R_4N)_n^{(4-n)-} + RV_{aq}^{2+} \longrightarrow \longrightarrow P(R_4N)_n^{4-n} + RV^+ \qquad (7)$$

$$P^*(R_4N)_n^{(4-n)-} + RV_{mic}^{2+} \longrightarrow P(R_4N)_n^{(3-n)-} + RV_{mic}^+ \qquad (8)$$

no long-lived transient ions but reaction 8 offers the possibility of transient charge separation, particularly if the RV_{mic}^+ is drawn to a more hydrophobic site upon reduction (37). The observation that quenching rates in SDS increase as more hydrophobic electrolytes are added could be consistent with either (or both) quenching processes 7 and 8 being of importance since the more hydrophobic cations should both increase the concentration of viologen in the aqueous phase as well as the concentration of associated porphyrin capable of interacting with the micelle. The fact that back reaction rates increase with increasing hydrophobicity of the added cation for MV^{2+} but decrease for the more hydrophobic BV^{2+} can be attributed to differences in hydrophobicity and hence binding capabilities for the reduced species MV^+ and BV^+. Thus for the former species addition of hydrophobic cations effectively decreases the sites available for binding of the reduced form and the back reaction rate increases; evidently BV^+ is sufficiently hydrophobic so that competition between R_4N^+ and BV^+ for sites is relatively less important.

For the situation where cationic detergent (CTAC) is present there is probably competition between the micelle and aqueous cations for the anionic porphyrin such that quenching takes place both at the micelle water interface and in the aqueous solution. Since CTAC is considerably more hydrophobic than SDS, it is reasonable to conclude that like-charge repulsion should be less important for the former while hydrophobic effects should be more significant. The back reaction rates decrease as the hydrophobicity of added cations increases and as the viologen becomes more hydrophobic. Although these trends could be attributed to a number of factors, we suspect that association of reduced viologen with the CTAC micelles not containing porphyrin may play a major role in slowing the back reaction in these cases. The effect of increased hydrophobicity of added R_4N^+ in this case would then be associated with desorption of porphyrin from the micelle.

Net Chemical Conversion via Selective Scavenging of Porphyrin-Viologen Photoredox Products in Micellar Systems

Since the palladium porphyrin-viologen-SDS system can give reasonable yields of redox products having relatively long lifetimes, it was of interest to explore possible subsequent reactions occurring with the redox radicals generated by photo-induced electron transfer, especially those in which both substrate and quencher could be regenerated concurrent with net chemical conversion. For the present system in which the oxidized porphyrin (P^{3-}) and reduced viologen ($MV^{\dot{+}}$) are generated, it is clear that selective reaction of either product should be possible. The reduced viologen is a moderate reductant and a number of previous studies have demonstrated that reactive colloidal redox catalysts can intercept $MV^{\dot{+}}$ and catalyze rapid reduction of water to hydrogen in neutral or acidic media (38-47). In flash photolysis experiments we find that addition of colloidal platinum-PVA results in a shortened lifetime for the portion of the transient spectrum corresponding to the reduced viologen with a concurrent increase in the lifetime of the P^{3-} transient. Unfortunately, the oxidized palladium porphyrin, a π-cation, is unstable and it decays relatively slowly ($k \sim 10^{-3}s^{-1}$) in aqueous SDS with some net decomposition of the porphyrin. Addition of iodide to the solution circumvents the degradation of the porphyrin as the iodide can be oxidized to triiodide (eqs. 9-11). A flash spectroscopic examination of solutions containing the palladium porphyrin,

$$P^{3-} + I^- \longrightarrow I\cdot + P^{4-} \qquad (9)$$

$$2I\cdot \longrightarrow I_2 \qquad (10)$$

$$I_2 + I^- \longrightarrow I_3^- \qquad (11)$$

MV^{2+}, SDS, and I^- but no catalyst shows a rapid recovery of the bleached porphyrin at 525 nm and a subsequent increase in absorption due to I_3^- ; no permanent change occurs in this case since the I_3^- oxidizes the reduced viologen. However, when platinum-PVA and I^- are simultaneously added to deaerated solutions containing palladium porphyrin, MV^{2+} and SDS, there is a net buildup of the I_3^--PVA complex which absorbs strongly at 490 nm ($\varepsilon \cong 42{,}000\ M^{-1}\ cm^{-1}$ for I_3^-) (48,49). Under these conditions there is little or no net decomposition of the porphyrin and it is evident that net cleavage of HI is occurring. Unfortunately the system is not practical since the net quantum yield is relatively low (∿0.001) and the extent of conversion obtained very small due to masking of the porphyrin absorbance by that of I_3^-.

A more interesting reaction in terms of overall quantum efficiency involves the use of iodide in conjunction with molecular oxygen to intercept selectively oxidized porphyrin and reduced viologen respectively as outlined in eqs. 12-19.

$$MV^{+} + O_2 \longrightarrow MV^{2+} + O_2^- \quad (12)$$

$$O_2^- + H^+ \longrightarrow HO_2\cdot \quad (13)$$

$$HO_2\cdot + I^- \longrightarrow I\cdot + HO_2^- \quad (14)$$

$$HO_2^- + H^+ \longrightarrow H_2O_2 \quad (15)$$

$$H_2O_2 + 2HI \longrightarrow 2H_2O + I_2 \quad (16)$$

The net reaction occurring here is given by eq. 17; although this

$$4HI + O_2 \longrightarrow 2I_2 + 2H_2O \quad (17)$$

reaction is clearly energetically downhill, it does not occur rapidly in the dark or in the absence of the porphyrin photocatalyst. The reaction is mechanistically interesting in that the protonation of superoxide (eq. 12) generates an oxidant from what was initially a reducing radical (MV^{+}); in this way the photoreaction can provide a limiting quantum yield of two for iodide oxidation; however, subsequent reaction of H_2O_2 with HI (eq. 15) (which is spontaneous!) indicates that the true limiting quantum yield is *four*. The reaction up to formation of H_2O_2 (eqs. 9-15) is an oxidative counterpart to the tertiary amine mediated photoreduction of polypyridyl ruthenium(II) complexes where deprotonation of an oxidized fragment provides the net generation of two reduced species/photon. In the present experiments with $[MV^{2+}]$ = 0.005 M, [NaI] = 0.05 M, [SDS] = 0.02 M and 1% PVA, we obtain $\phi_{I_3^-}^{405\ nm} = 0.30 \pm 0.05$; since each I_3^- comes from two oxidized iodide ions the net efficiency of iodide oxidation is 0.60 or about 15% of the theoretical reaction quantum efficiency. Here again the reaction cannot be driven to large conversions because of competitive absorption by the product I_3^- complex. However, the moderately high overall initial efficiency for a reaction initiated by an excited state electron transfer process in which both substrate and quencher are recycled is noteworthy and suggests that other useful applications in similar reactions can be forthcoming.

The Use of Extramicellar Probe Luminescence Quenching to Monitor Binding of Charged Substrates to Micelles

The studies described up to this point have focused on the use of micelles to modify the rates and net processes occurring in reaction sequences initiated by light-induced electron transfer. The remainder of the paper will deal with studies in which the same basic reaction is used to probe the structure and binding properties of anionic (SDS) micelles. As pointed out earlier, in many cases it is possible to "tune" many of the properties of transition metal complexes by varying ligand-substituents. One example where the chief effect is on solubility and hydrophilicity involves the use of the dianions, 4,4'-dicarboxy-2,2'-bipyridine and 5,5'-dicarboxy-2,2'-bipyridine to form anionic ruthenium complexes, RuL_3^{4-}, 5 and 6, which are water soluble and highly hydrophilic. These complexes are luminescent in aqueous solution; the luminescence of 5 and 6 is strongly quenched by addition of MV^{2+}. In the case of 5 the quenching in water gives coincident lifetime and intensity Stern-Volmer plots (Figure 1) with added

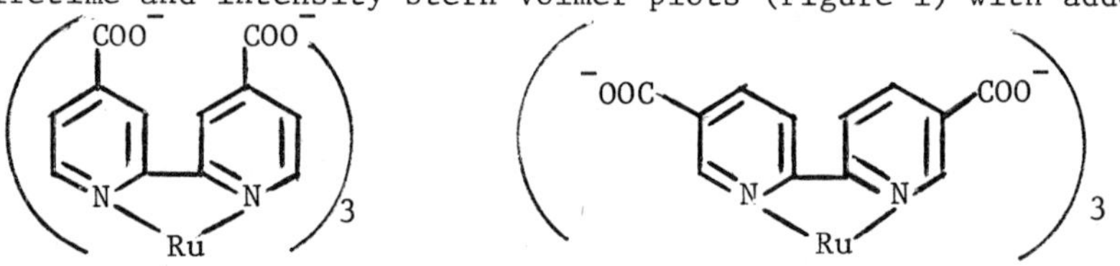

COO⁻ COO⁻ N N Ru 3
5

⁻OOC COO⁻ N N Ru 3
6

NaCl above 0.1 M; at lower salt concentrations the intensity plots show upwards curvature indicating a slight amount of association between 5 and the quencher. The dynamic quenching constant obtained is $5.6 \times 10^9\ M^{-1}\ s^{-1}$. A similar study with Cu^{2+} as a quencher leads to linear lifetime Stern-Volmer plots giving a quenching constant, $k_q = 5.8 \times 10^9\ M^{-1}\ s^{-1}$

Addition of SDS to aqueous solutions above the cmc leads to attenuated quenching for both cations as would be expected due to binding of the cations to the anionic micelles. The behavior observed differs sharply for the two quenchers, however,(Figure 1) suggesting important differences in the binding of the cations to the micelle. The behavior for MV^{2+} is particularly striking and clearcut: addition of SDS above the cmc results in Stern-Volmer plots showing two distinct linear regions (Figure 1). The quenching is strongly attenuated until a certain $[MV^{2+}]$, directly related to the [SDS], is attained; at this point a sharp break occurs and the enhanced quenching above this point gives a plot parallel to that obtained with no quencher. The ratio of $[SDS]_m/[MV^{2+}]$ at the "break point" is 3.3 ± 0.2. For Cu^{2+} the attenua-

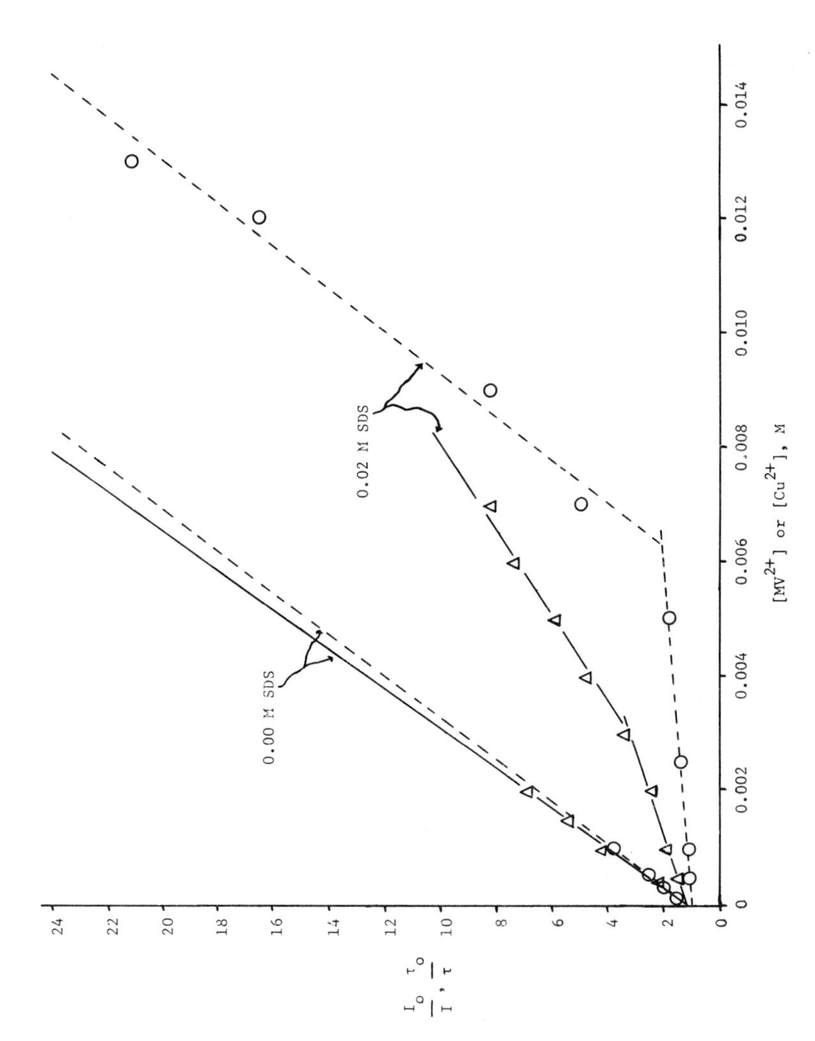

Figure 1. Luminescence quenching of compound 5 by MV^{2+} or Cu^{2+} at various SDS concentrations with 0.05 M NaCl added as electrolyte. I_o/I points (○) are for MV^{2+}; τ_o/τ points (△) are for Cu^{2+}.

tion of quenching is less pronounced and instead of two distinct linear regions a much more curved plot is obtained (Figure 1). Attempts to extrapolate to a "break point" analogous to that clearly indicated on the MV^{2+} plots give a value of $[SDS]_m/[Cu^{2+}] = 5.9 \pm 0.4$ (50).

The rather striking differences in behavior observed for the two dications as quenchers of the luminescence of 5 can be interpreted in terms of different modes of binding to the anionic micelle. For Cu^{2+} the attractive forces leading to binding to the SDS micelle are primarily electrostatic and probably involve close association of the Cu^{2+} to the anionic headgroups. The curvature in the quenching plot with Cu^{2+} (Figure 1) is attributed to a reduced tendency of the micelle to bind additional Cu^{2+} ions as the $[Cu^{2+}]$ increases due to screening of the micellar charge, as increasing numbers of Cu^{2+} ions associate with the headgroups. The binding capacity estimated from the plot—ca. one Cu^{2+} for every six surfactant molecules in the micelle—is very close to the binding capacities measured for Cu^{2+} and SDS using quite different methods (51-53), and compares well with that measured for other divalent metal ions such as Ni^{2+} and Mn^{2+} (51, 52, 54).

The results obtained for MV^{2+} suggest markedly different binding behavior toward the SDS micelle. Here the observation of little upwards curvature in the attenuated quenching region suggests that binding of several viologen molecules to the micelle does not reduce the affinity of the micelle for additional viologens until a limiting capacity is reached. The limiting capacity is almost twice that observed for the much smaller metal ions; the differences observed clearly suggest that the viologen is binding in sites different from those occupied by Cu^{2+} (and other metal ions) and that different factors govern the binding. Behavior similar to that obtained with MV^{2+} has been observed for other organic cations such as BV^{2+} and 4-cyano-N-benzyl-pyridinium (CBP^+) (50). The behavior observed with CBP^+ is particularly striking since the binding capacity estimated from the break in the quenching plot is one CBP^+ for every two SDS molecules in the micelle. In contrast, the less hydrophobic cation 4-cyano-N-methylpyridinium (CMP^+) shows relatively small attenuation in its quenching behavior indicating its affinity for the micelle is much reduced. We interpret the results of these experiments to indicate that the organic cations bind to sites away from the headgroups where there would otherwise be a hydrocarbon-water interface. These regions are perhaps best described as hydrophobic-hydrophilic sites in which the insertion of a moderately polar solute (in these cases cations) between hydrocarbon and water eliminates a high energy interface. This can lead to micelle solubilization or binding of reagents which are not particularly soluble in either pure water or alkane solvent alone. The observation that the capacity of the micelle to absorb very large numbers of these ions is in accord with increasing evidence (55-58) that there is a large surface-volume ratio for micelles and, perhaps more importantly, a large hydrocarbon-water interface region. These results also agree with recent models suggesting a considerably more open structure for simple micelles than conventional models in which there is a sharp polar-nonpolar boundary.

Acknowledgement: We are grateful to the National Institutes of Health (Grant GM15,238) and National Science Foundation (Grant #CHE 7823126) for support of this work. J. Bonitha thanks the Fundação de Amparo à Pesquisa do Estado de São Paulo (FAPESP 80/0232 to J. B. S. Bonilha).

Literature Cited

1. Sutin, N. J. Photochem. 1979, 10, 19-36.
2. Demas, J.N.; and Crosby, G.A. J. Am. Chem. Soc. 1970, 92, 7262-7270.
3. Gafney, H.D.; Adamson, A.W. J. Am. Chem. Soc.1972, 94, 8238-8239.
4. Bock, C.R.; Meyer, T.J.; Whitten, D.G. ibid. 1974, 96, 4710-4712.
5. Bock, C.R.; Meyer, T.J.; Whitten, D.G. ibid. 1975, 97, 2909-2911.
6. Navon, G.; Sutin, N. Inorg. Chem.1974, 13, 2159-2164.
7. Lawrence, G.S.; Balzani, V. ibid. 1974, 13, 2976-2982.
8. Young, R.C.; Meyer, T.J.; Whitten, D.G. J. Am. Chem. Soc., 1975, 97, 4781-4782.
9. Young, R.C.; Meyer, T.J.; Whitten, D.G. ibid. 1976, 98, 286-287.
10. Creutz, C.; Sutin, N. Inorg. Chem. 1976, 15, 496-498.
11. Lin, C.T.; Bottcher, W.; Chou, M.; Creutz, C.; Sutin, N. J. Am. Chem. Soc. 1976, 98, 6536-6544.
12. Lin, C.T.; Sutin, N. ibid. 1975, 97, 3543-3545.
13. Creutz, C.; Sutin, N. ibid. 1977, 99, 241-243.
14. Toma, H.E.; Creutz, C. Inorg. Chem. 1977, 16, 545-550.
15. Balzani, V.; Moggi, L.; Manfrin, M.F.; Bolletta, F.; Lawrence, G.S. Coord Chem. Revs. 1975, 15, 321-433.
16. Juris, A.; Gandolfi, M.T.; Manfrin, M.F.; Balzani, V. J. Am. Chem. Soc. 1976, 98, 1047-1048.
17. Sutin, N.; Creutz, C. Adv. Chem. Ser. 1978, 168, 1-27.
18. DeLaive, P.J.; Lee, J.T.; Abrũna, H.; Sprintschnik, H.W.; Meyer, T.J.; Whitten, D.G. Adv. Chem. Ser. 1978, 168, 28-43.
19. DeLaive, P.J.; Giannotti, C.; Whitten, D.G. Adv. Chem. Ser. 1979, 173, 236-251.
20. Ballardini, R.; Varani, G.; Indelli , M.T.; Scandola, F.; Balzani, V. J. Am. Chem. Soc. 1978, 100, 7219-7223.
21. Whitten, D.G. Accounts Chem. Res. 1980, 13, 83-90.
22. (a) Thomas, J.K. Acc. Chem. Res. 1977, 10, 133; (b) (b) Kalyanasundaram, K. Chem. Soc. Revs. 1978, 4, 453-472.
23. (a) Burrows, H.D.; Formosinho, S.J.; Pawa, F.J.R. J. Photochem. 1980, 12, 285-292; (b) Russell, J.C.; Braun, A.M.; Whitten, D.G.; unpublished results.

24. (a) Moroi, Y.; Infelta, P.P.; Grätzel, M. J. Am. Chem. Soc. 1979, 101, 573-577; (b) Moroi, Y.; Braun, A.M.; Grätzel, M. J. Am. Chem. Soc. 1979, 101, 567-573.
25. Meisel, D.; Matheson, M.S.; Rabani, J. J. Am. Chem. Soc. 1978, 100, 117-122.
26. (a) Rodgers, M.A.J.; Wheeler, M.F.; da Silva, E. Chem. Phys. Lett. 1978, 53, 165-169; (b) Rodgers, M.A.J.; Wheeler, M.F.; da Silva, E. Chem. Phys. Lett. 1976, 43, 587-591.
27. Turro, N.J.; Grätzel, M.; Braun, A.M. Angew Chem. Int'l Ed. Engl. 1980, 19, 549-675.
28. Schmehl, R.H.; Whitten, D.G. J. Am. Chem. Soc. 1980, 102. 1938-1941.
29. (a) Infelta, P.P.; Grätzel, M.; Fendler, J.H. J. Am. Chem. Soc. 1980, 102, 1479-1484; (b) Ford, W.E.; Otvos, J.W.; Calvin, M. Proc. Nat. Acad. Sci. U.S.A.1979, 76, 3590-3593, and references therein. (c) Sudo, Y.; Toda, F. Nature 1979, 279, 807-809.
30. (a) Tabushi, I.; Funakura, M. J. Am. Chem. Soc. 1976, 98, 4684-4685; (b) Tien, H.T. Nature 1968, 219, 272-274; (c) Hesketh, T.R. Nature 1969, 224, 1026-1028; (d) see Photochem. Photobiol 1976, 24, 12; (e) Nekrasov, L.I.; Chasovnikova, L.V.; Kobozen, N.I. J. Phys. Chem. USSR 1967, 41, 1426-1430.
31. Tsutsui, Y.; Takuma, K.; Nishijuma, T.; Matsuo, T. Chem. Lett. 1979, 617-620.
32. Razem-Katusin, B.; Wong, M.; Thomas J.K. J. Am. Chem. Soc. 1978, 100, 1679-1687.
33. (a) Moroi, Y.; Infelta, P.P.; Grätzel, M. J. Am. Chem. Soc. 1979, 101, 573-577; (b) Moroi, Y.; Braun, A.M.; Grätzel, M. J. Am. Chem. Soc. 1979, 101, 567-573.
34. Brugger, P.A.; Grätzel, M. J. Am. Chem. Soc. 1980, 102, 2461-2463.
35. Rodgers, M.A.J.; Becker, J.C. J. Phys. Chem. 1980, 84, 2762.
36. Schmehl, R.H.; Whitesell, L.G.; Whitten, D.G. J. Am Chem. Soc. in press.
37. Brugger, P.-A; Infelta, P.P.; Braun, A.M.; Grätzel, M. J. Am. Chem. Soc. 1981, 103, 320.
38. Lehn, J.-M.; Sauvage, J.-P. Nouv. J. Chem. 1977, 1, 449-451.
39. Kalyanasundarum, K.; Kiwi, J.; Grätzel, M. Helv. Chim. Acta. 1978, 61, 2720-2730.
40. Moradpour, A.; Amouyal, A.; Keller, P.; Kagan, H. Nouv. J. Chem. 1978, 2, 547-549.
41. Koryakin, B.V.; Dzabiev, T.S.; Shilov, A.E. Dokl. Akad. Nauk. SSSR 1976, 229, 614-620.
42. Koryakin, B.V.; Kzabiev, T.S.; Shilov, A.E. Dokl. Akad. Nauk. SSSR 1976, 238, 620-625.
43. Lehn, J.-M.; Sauvage, J.-P.; Ziessel, R. Nouv. J. Chim. 1979, 3, 423-427.
44. Kiwi, J.; Bogarello, E.; Pelizzetti, E.; Visca, M.; Grätzel, M. Angew. Chem. Int. Ed. Engl. 1980, 19, 646-648.

45. Kalyanasundarum, K.; Grätzel, M. Angew Chem. Int. Ed, Engl. 1979, 18, 701.
46. Lehn, J.-M.; Sauvage, J.-P.; Ziessel, R. Nouv. J. Chim. 1980, 4, 355-358.
47. Keller, P.; Moradpour, A. J. Am. Chem. Soc. 1980, 102, 7193-7196.
48. Tebelev, L.G.; Silkina, N.A. Dokl Akad.Nauk. SSSR 1965, 161, 1096-1098.
49. Mokhnach, V.O.; Zueva, I.L. Dokl. Akad. Nauk. SSSR 1961, 136, 832-834.
50. Foreman, T.K.; Sobol, W.M.; Whitten, D.G., submitted for publication.
51. Fischer, M.; Knoche, W.; Fletcher, P.D.I.; Robinson, B.H.; White, N.C. Colloid and Polymer Science 1980, 258, 733.
52. Grätzel, M.; Thomas, J.K. J. Phys. Chem. 1974, 78, 2248.
53. Grieser, F.; Tausch-Treml, R. J. Am. Chem. Soc. 1980, 102, 7258.
54. Oakes, J.J.; Chem. Soc., Faraday Trans. 2 1973, 69. 1321.
55. Mukerjee, P.; Cardinal, J.R. J. Phys. Chem. 1978, 82, 1620.
56. Cardinal, J.R.; Mukerjee, P. J. Phys. Chem. 1978, 82, 1614.
57. Menger, F.M. Accounts Chem. Res. 1979, 12, 111.
58. Mukerjee, P. "Solution Chemistry of Surfactants", Mittal, K.L., ed.; Plenum: New York, 1979, pp 153-174.

RECEIVED May 27, 1981.

4

Aspects of Artificial Photosynthesis

The Role of Potential Gradients in Promoting Charge Separation in the Presence of Surfactant Vesicles

MOHAMMAD S. TUNULI and JANOS H. FENDLER

Texas A&M University, Department of Chemistry, College Station, TX 77843

Completely synthetic vesicles, prepared from dioctadecyldimethylammonium chloride, DODAC, and dihexadecylphosphate, DHP, have been used in our laboratories as media for investigating aspects of artificial photosynthesis over the past several years. Different potentials, created by the high charge densities on the surface of DODAC and DHP vesicles, and their exploitation for photochemical solar energy conversion are discussed in this presentation. Surface potential, charge separation potential, diffusion potential and Donnan potential are exploited for enhanced energy and electron transfer on charged vesicle surfaces, for the utilization of field effects for charge separation, for partitioning between the inner and outer compartments of radicals expelled from vesicle bilayers and for facilitating electron transfer across bilayers.

Photochemical solar energy conversion is a vitally important and extremely active area of research (1-9). The excited state of a suitable sensitizer, produced by irradiation, is a better electron acceptor as well as a better electron donor than its ground state. Light absorption can drive, therefore, a redox reaction nonspontaneously and result in the storage of energy, ΔG, in D^+ and A^-:

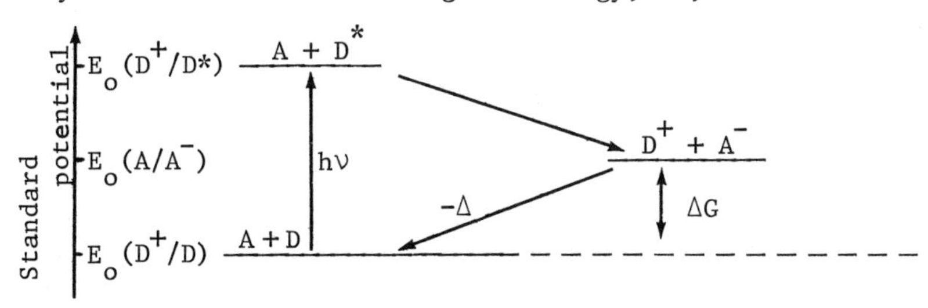

0097-6156/82/0177-0053$05.00/0

Further, if D^+ and A^- have appropriate redox potentials they may directly reduce water to hydrogen and oxidize it to oxygen:

$$2A^- + 2H_2O \longrightarrow 2A + 2OH^- + H_2\uparrow \quad (2\ e^-\ \text{reduction}) \tag{1}$$

$$4D^+ + 2H_2O \longrightarrow 4D + 4H^+ + O_2\uparrow \quad (4\ e^-\ \text{oxidation}) \tag{2}$$

Unfortunately, in homogeneous solution the rapid recombination of $D^+ + A^-$ to D + A precludes the possibility of this type of photochemical energy storage and conversion. Separation of charges in space provides an efficient method for diminishing undesirable charge recombinations. Aqueous (10) and reversed (11,12) micelles, microemulsions (13), monolayers (14), bilayer (black) lipid membranes (15), polyelectrolytes (16), liposomes (17) and surfactant vesicles (18-27) have been used to affect charge separation by organizing sensitizers, electron donors and acceptors in their compartments (28). These organized assemblies are expected to (a) solubilize, concentrate, compartmentalize, organize, and localize reactants; (b) maintain proton and/or reactant gradients; (c) alter quantum efficiencies; (d) lower ionization potentials; (e) change oxidation and reduction properties; (f) change dissociation constants; (g) affect vectorial electron displacements; (h) alter photophysical pathways and rates; (i) alter chemical pathways and rates; (j) stabilize reactants and/or products and/or transition states; (k) separate charges and/or products; and (l) be chemically stable, optically transparent and photochemically inactive.

Completely synthetic surfactant vesicles have been used in our laboratories as media for examining aspects of artificial photosynthesis. To-date, we have investigated vesicles formed from dioctadecyldimethylammonium chloride, DODAC, and dihexadecylphosphate, DHP (19-21, 23, 27):

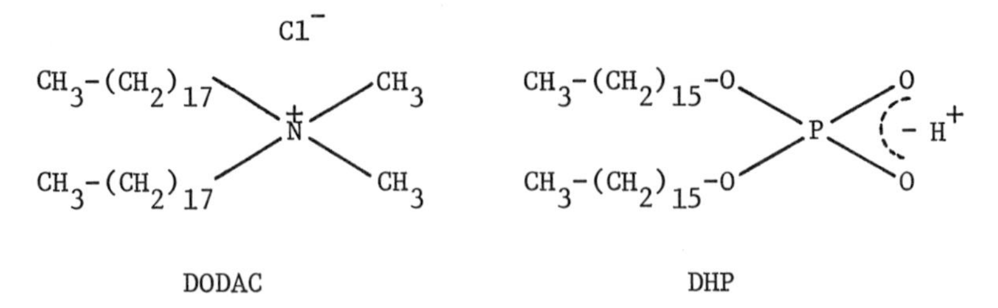

Dispersal of DODAC or DHP in water by ultrasonic irradiation results in the formation of fairly uniform single compartment vesicles (Figure 1). DODAC and DHP vesicles are stable for weeks

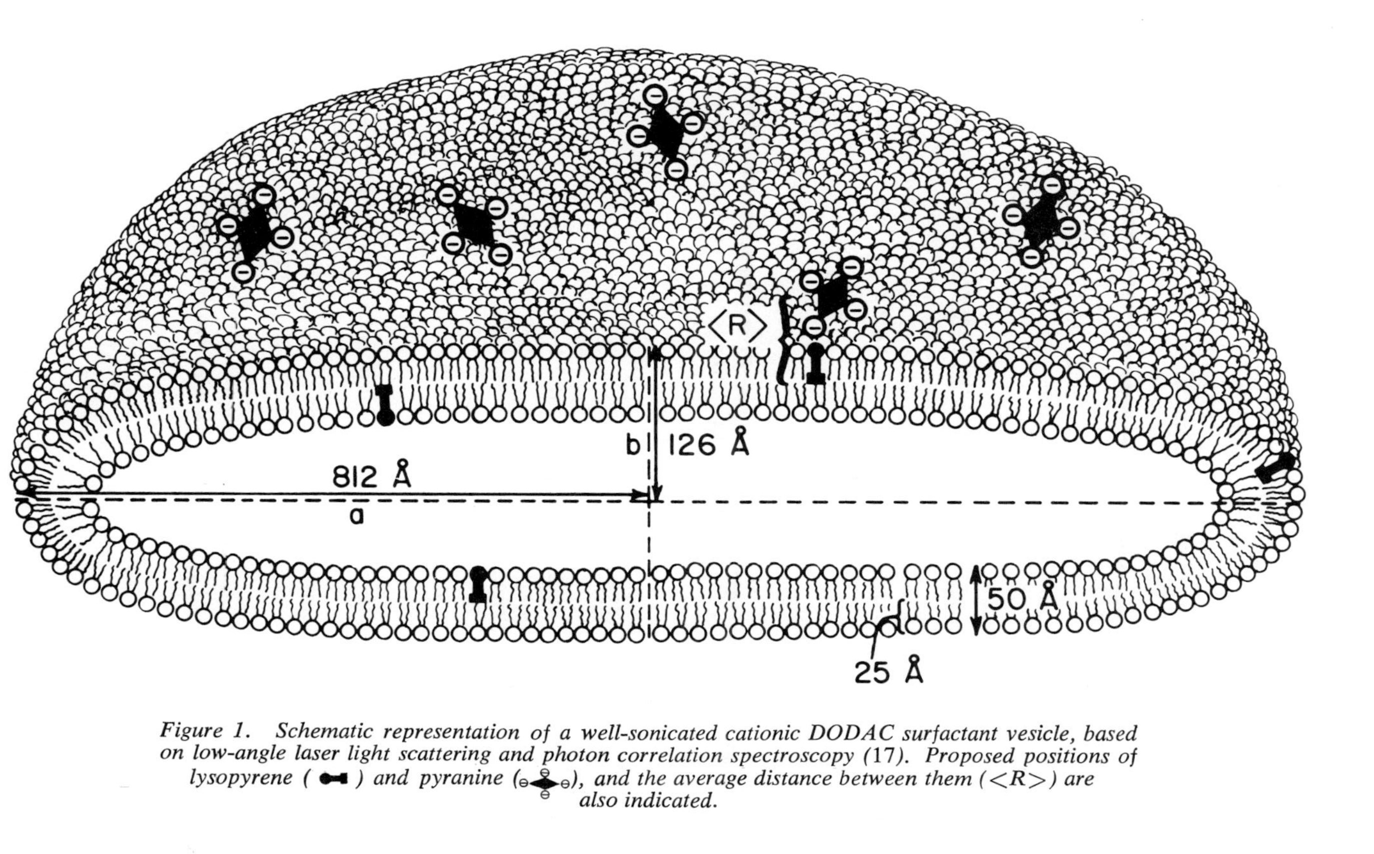

Figure 1. Schematic representation of a well-sonicated cationic DODAC surfactant vesicle, based on low-angle laser light scattering and photon correlation spectroscopy (17). Proposed positions of lysopyrene ( ) and pyranine ( ), and the average distance between them (<R>) are also indicated.

in pH 2-12 range, osmotically active, undergo thermotropic phase transitions and, most importantly entrap and retain molecules in their compartments. Advantages of surfactant vesicles over other systems are that they are able to organize large numbers of sensitizers, electron donors and acceptors per aggregate and that they are amenable to electrostatic modification and chemical functionalization. Importantly, unlike natural membranes which are composed mostly of zwitterionic lipids, surfactant vesicles are highly charged and have high charge densities on their surfaces. These charges create appreciable potentials. The types of potentials associated with surfactant vesicles and their exploitation in photochemical solar energy conversion are the subject of this presentation.

Types of Potentials Associated with Surfactant Vesicles

In addition to the surface potential, Ψ_o, present at the outer and inner surfaces of charged vesicles several additional potentials can be created. Of these, the charge separation potential, the diffusion potential and the Donnan potential will be briefly discussed.

Surface Potential. The presence of ionized head groups on a spherical vesicle with radius r and charge q_v on the vesicle leads to a surface potential Ψ_o:

$$\Psi_o = q_v e^2/\varepsilon r \tag{3}$$

where ε is the dielectric constant at the interface. The surface potential decreases with increasing distances from the charged surface. At a distance x, from the surface the potential is given by (see Figure 2):

$$\Psi_x = \Psi_o \exp(-\beta x) \tag{4}$$

where β = (distance from surface)/(distance from the outer Helmholtz plane).

In the Gouy-Chapman diffuse layer the concentration-distance profile is given by the Boltzmann distribution:

$$C_x = C_o \exp\left(-\frac{z_i F \Psi_x}{RT}\right) \tag{5}$$

where C_x and C_o are the concentration of ions at distance x from the surface and at the bulk, respectively, z_i is the number of units of electronic charge on ion i, F and R are the Faraday and universal gas constants and T is the absolute temperature. The Poisson equation relates the potential profile to the charge density by:

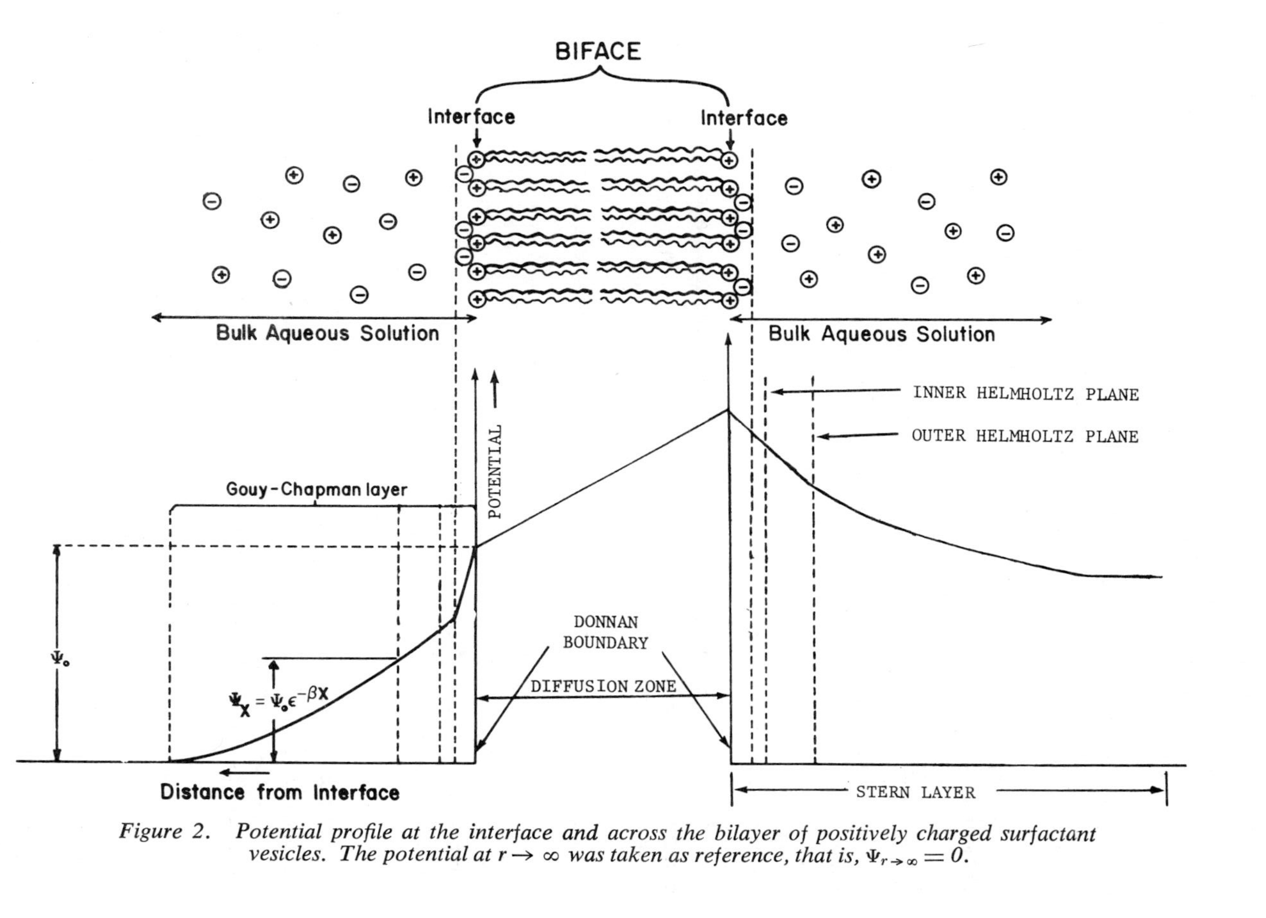

Figure 2. Potential profile at the interface and across the bilayer of positively charged surfactant vesicles. The potential at r → ∞ was taken as reference, that is, $\Psi_{r\to\infty} = 0$.

$$\frac{d^2\Psi}{dx^2} = -\frac{F}{\varepsilon}\sum_i z_i C_o \exp\left(-\frac{z_i F \Psi_x}{RT}\right) \tag{6}$$

which upon integration (using appropriate boundary conditions) gives an expression for the electric field:

$$X_x = \pm\left\{\frac{2RT}{\varepsilon}\sum_i C_o\left[\exp\left(-\frac{z_i F\Psi_x}{RT}\right) - 1\right]\right\}^{1/2} \tag{7}$$

<u>Charge Separation Potential</u>. An electron can be transferred across the bilayer of surfactant vesicles from a donor to an acceptor. This transfer renders the donor side of the vesicle surface to be more positive than the acceptor side, thus it creates a potential, referred to as charge separation potential, $\Psi_{c.s}$. This, by analogy to charging a parallel plate condenser, is given by:

$$\Psi_{c.s} = \frac{q'_m \delta}{A\varepsilon\varepsilon'} \tag{8}$$

where q'_m is the additional charge deposited on the vesicle due to the photoinitiated electron transfer, δ is the thickness of the surfactant vesicle, A is its surface area, ε denotes the dielectric constant of the interfacial region and ε' is the permittivity of free space. If $\Psi_{c.s}$ is known, q'_m can be obtained from:

$$q'_m = \pm\left\{2RT\varepsilon\sum_i C_{i,o}\left[\exp\left(-\frac{z_i F\Psi_{c.s}}{RT}\right) - 1\right]\right\}^{1/2} \tag{9}$$

where $C_{i,o}$ is the bulk concentration of electron donor/acceptor. Negative sign is used if $\Psi_{c.s}$ is positive and *vice versa*.

<u>Diffusion Potential</u>. Diffusion potential arises from a concentration gradient, ∇C_i, across the vesicle bilayer. Addition of an electrolyte whose component ions differ in their mobilities to already formed vesicles will give rise to spacial segregation of ions. Sodium chloride provides an example for such a behavior. At 298.15°K the mobilities of Na^+ and Cl^- are 5.2×10^{-4} $cm^2 sec^{-1} v^{-1}$ and 7.9×10^{-4} $cm^2 sec^{-1} v^{-1}$.

For a vesicle of thickness δ, immersed in a stagnant (i.e. no convection) dilute electrolyte solution the transport of species across the bilayer can be formulated by the following set of equations (<u>29</u>):

$$J_i = \underbrace{-z_i \mu_i F C_i \nabla\Psi}_{\text{migration}} - \underbrace{D_i \nabla C_i}_{\text{diffusion}} \tag{10}$$

$$\frac{dC_i}{dt} = -\nabla J_i \qquad (11)$$

$$i \text{ (ion current)} = F\sum_i z_i J_i \qquad (12)$$

$$\sum_i z_i c_i = o \text{ (principle of electroneutrality)} \qquad (13)$$

where J_i is the ionic flux under the joint influence of electrical potential and concentration gradients, μ_i, is the mobility of the ith species, and D_i is the diffusion coefficient of the ith species. Equation 10 can be integrated numerically without much difficulty. A simpler approach has been provided by Goldman who assumed the constancy of the field across the membrane (30). The problem then becomes analogous to that of electric conduction in the copper - - copper oxide rectifier (31). Using this assumption, integration of equation 10 leads to

$$J_i^{\pm} = \frac{\mu_i^{\pm} F}{\delta}\,\Psi_d\,\frac{a_o^{\pm} - a_i^{\pm}\exp\dfrac{z_i^{\pm}F\Psi_d}{RT}}{1 - \exp\dfrac{z_i^{\pm}F\Psi_d}{RT}} \qquad (14)$$

where Ψ_d is the diffusion potential, $a_o^{\pm}$ and $a_i^{\pm}$ are the activity coefficients and μ_i is the mobility of the species involved.

Donnan Potential. Coion exclusion and counterion condensation on the charged vesicle surface creates the well known Donnan potential. The Donnan potential can be derived either by a kinetic or by a thermodynamic approach (32). Using the kinetic approach, the mass transport equation is written by:

$$J_i = - D_i C_i (\nabla \ln a_i + \frac{z_i F}{RT}\nabla\Psi) + C_i(1 - \sigma_i)J_v \qquad (15)$$

where the flux, J_i, if directed toward the vesicle, is considered to be positive. C_i is the concentration of the species in the interfacial region, J_v is the volume flux through the vesicle, σ_i the "reflection" coefficient, represents some specific interaction of the species with vesicle. Since at the interface of charged vesicles both diffusion and migration fluxes are much larger than the flux due to convection, equation 16 can be written:

$$\frac{z_i F}{RT}\nabla\Psi = - \nabla \ln a_i \qquad (16)$$

which upon integration gives the Donnan potential:

$$\Psi_{DONNAN} = \overline{\Psi} - \Psi_{solution} = -\frac{RT}{z_i F}\ln\frac{\overline{ai}}{ai}$$
$$\Psi_{DONNAN} = -\frac{RT}{z_i F}\ln\frac{\overline{m}_i\overline{\gamma}_i}{m_i\gamma_i} \tag{17}$$

where m and γ represent molal concentrations and activity coefficients and the bar refers to the Donnan phase. Concentrations of the counterions accumulated in the Donnan phase are given by:

$$(\overline{a}_i)_\pm = \sqrt{\frac{\overline{M}^2}{2} + \frac{a_\pm^2}{\overline{\gamma}_\pm^2}} \pm \frac{\overline{M}}{2} \tag{18}$$

where $\overline{M}$ is the concentration of fixed charges on the vesicle and $a_\pm^2$ represents the mean ion activity coefficients ($a_\pm^2 = \overline{m}_+\overline{\gamma}_+\overline{m}_-\overline{\gamma}_-$)

Exploitation of Potentials for Energy and Electron Transfer and Charge Separation in Surfactant Vesicles

Enhanced Energy and Electron Transfer on Charged Vesicle Surfaces. Localization of molecules in biological matrices is an essential requirement for many processes. Energy transfer *in vivo* photosynthesis is largely dependent, for example, on the precise location of chlorophyll molecules in the chloroplast (33). An average distance of approximately 15 Å between chlorophylls is considered to be ideal for efficient energy transfer without self-quenching. Ionic surfactant vesicles attract oppositely charged species onto their surfaces. Intravesicular energy and electron transfer readily occur in the potential field of the aggregate at reduced dimensionalities (34).

Efficient intramolecular energy transfer has been observed in DODAC vesicles (20). The donor, 2-hydroxy-1-[ω-(1-pyrene)decanoyl]--sn-glycero-3-phosphatidylcholine (lysopyrene), was localized in the hydrophobic bilayer of the vesicles. The acceptor, trisodium 8-hydroxy-1,3,6-pyrenetrisulfonate (pyranine), having four negative charges, was attracted to the outer surface of positively charged DODAC vesicles (Figure 1). Depending on the concentration of pyranine, energy transfer efficiencies up to 43% have been observed (20). Conversely, energy transfer efficiencies in the absence of vesicles were less than 3%. The apparent rate constant for energy transfer quenching, 6.2×10^{11} $M^{-1}sec^{-1}$, is the consequence of an approximate 1000-fold increase of acceptor concentration on the vesicle surface.

Efficient photosensitized electron transfer has also been observed from tris(2,2'-bipyridine)ruthenium cation $Ru(bpy)_3^{2+*}$ to methylviologen:

$$Ru(bpy)_3^{2+} \xrightarrow{h\nu} Ru(bpy)_3^{2+*} \qquad (19)$$

$$Ru(bpy)_3^{2+*} + MV^{2+} \longrightarrow Ru(bpy)_3^{3+} + MV^{+\cdot} \qquad (20)$$

on the outer and inner surfaces of anionic DHP surfactant vesicles (Systems III and IV, respectively in Figure 3) (23). The apparent rate constant for reaction 20, $(4\text{-}5)10^{11}$ $M^{-1}sec^{-1}$, in Systems III and IV are three orders of magnitude greater than that found in water ($2x10^8$ $M^{-1}sec^{-1}$). Electron transfer is likely to occur by diffusion or hopping on the vesicle surface. Under typical conditions, approximately 60 molecules of $Ru(bpy)_3^{2+}$ and 300 molecules of MV^{2+} associate with each DHP vesicle. Taking charge repulsions into consideration, average areas for $Ru(bpy)_3^{2+}$ and MV^{2+} molecules are estimated to be 400 $Å^2$ and 200 $Å^2$, respectively. Since the surface area of a DHP vesicle is $1.2x10^7$ $Å^2$ (18) the maximum area the reactive partners need to cover prior to collision is only 200 $Å^2$. This value is orders of magnitude smaller than 10^5 $Å^2$ estimated for $\bar{d}^2$, the square of the mean diffusive displacement of $Ru(bpy)_3^{2+}$ and MV^{2+} (12). Thus, the reactive partners can readily find each other on the surface of the DHP vesicles within their lifetimes. Unfortunately, the close proximity also results in much enhanced back reaction:

$$Ru(bpy)_3^{3+} + MV^{+\cdot} \longrightarrow Ru(bpy)_3^{2+} + MV^{2+} \qquad (21)$$

Different organization is needed, therefore, to accomplish the desired efficiency in energy conversion; i.e., to enhance the rate of the forward electron transfer (reaction 20, for example) and at the same time reduce the back reaction (reaction 21, for example). Exploitation of potentials to accomplish this goal will be illustrated in the following sections.

Influence of Field Effect. Since electron transfer rates are directly related to the field, a judicious manipulation of the distance of a sensitizer and an electron acceptor (or donor) from a highly charged surface across the Stern layer (Figure 2, equation 7) is expected to result in altered efficiencies. This expectation has been realized in achieving effective charge separation under the influence of a positive electric field, generated by DODAC vesicles (35). Rate constant for electron transfer from L-cysteine to the excited state of $Ru(bpy)_3^{2+}$:

$$Ru(bpy)_3^{2+*} + \text{L-cysteine} \longrightarrow Ru(bpy)_3^{+} + \text{L-cystine}^{+} \qquad (22)$$

has been determined by laser flash photolysis (35). Satisfactory agreement has been obtained between the experimentally observed rate constants, k_{obs}^{exp}, and those calculated, k_{obs}^{calc}, on the basis of the presence of an electric field (Table I).

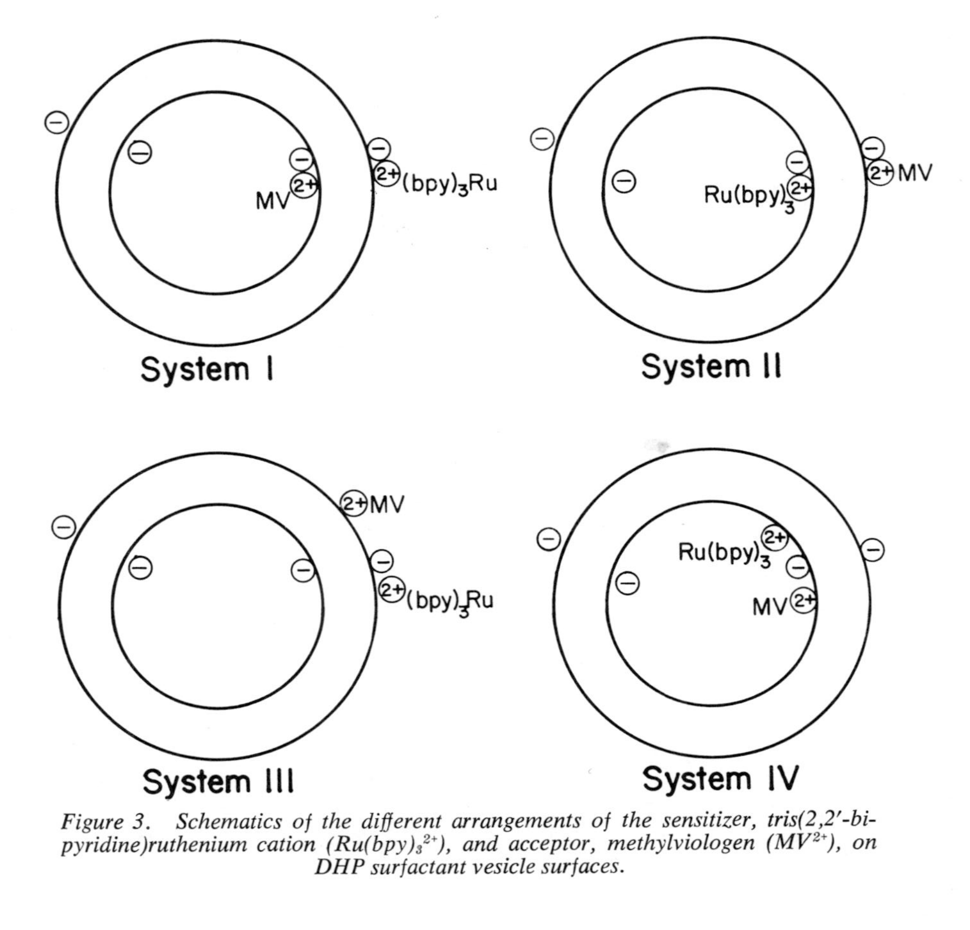

Figure 3. Schematics of the different arrangements of the sensitizer, tris(2,2′-bipyridine)ruthenium cation ($Ru(bpy)_3^{2+}$), and acceptor, methylviologen (MV^{2+}), on DHP surfactant vesicle surfaces.

TABLE I

Calculated and Observed Rate Constants for the Electron Transfer $Ru(bpy)_3^{2+*}$ + L-cysteine → $Ru(bpy)_3^{+}$ + L-cysteine^{+} in DODAC Vesicles

pH	k_{obs}^{exp}, $M^{-1}sec^{-1}$	k_{obs}^{calc}, $M^{-1}sec^{-1}$	calc.[a] ΔG, Kcal mole^{-1}	calc.[a] $\Delta G^{\ddagger}$, Kcal mole^{-1}
3.0	$3.4x10^7$	$6.7x10^7$	-25.8	1.4
8.4	$8.2x10^7$	$1.3x10^8$	-36.4	1.0
11.5	$1.2x10^8$	$1.7x10^8$	-44.3	0.8

[a] Calculated by means of equation 31. See the ensuing discussion for details on the calculations.

Details of obtaining the k_{obs}^{calc} values are as follows. The field operates through L-cysteine, whose charge and hence whose distance from the DODAC vesicle surface can be altered by changing the bulk hydrogen ion concentration of the solution. Dissociation constant of L-cysteine in the presence of surfactant vesicles, pKa, is given by:

$$pKa = pKa° - zF\Psi_x \tag{23}$$

where pKa° is the dissociation constant in water and Ψ_x is defined by equation 4. The redox potential for this process can be written as:

$$E_{D^+/D} = E^{\circ}_{D^+/D} - 0.059\left\{\log \frac{[HA^z]}{[A^{z-1}]} - 2\alpha(z-1)\sqrt{\mu} - - pKa° + \frac{zF\Psi_x}{2.3RT}\right\} \tag{24}$$

where α is a constant (0.509x2.303) and μ is the ionic strength, respectively. The free energy change for reaction 22 is given by:

$$\Delta G = -23.06\left[E_{Ru(bpy)_3^{2+*}/Ru(bpy)^+} - E_{D^+/D}\right] \tag{25}$$

and that for the rate by:

$$k_{et} = A\exp(-\Delta G^{\ddagger}/RT) \tag{26}$$

Assuming that the mechanism involves three steps; the diffusion of the reacting partners:

$$Ru(bpy)_3^{2+*} + \text{L-cysteine} \underset{k_{-D}}{\overset{k_D}{\rightleftharpoons}} Ru(bpy)_3^{2+*} \mid \text{L-cysteine} \tag{27}$$

electron transfer:

$$Ru(bpy)_3^{2+*} \mid \text{L-cysteine} \underset{k_{-et}}{\overset{k_{et}}{\rightleftharpoons}} Ru(bpy)^{+} \mid \text{L-cysteine} \tag{28}$$

charge separation:

$$Ru(bpy)_3^{+} \mid \text{L-cysteine}^{+} \xrightarrow{k_s} Ru(bpy)_3^{+} \mid \text{L-cysteine}^{+} \tag{29}$$

and thermal recombination:

$$Ru(bpy)_3^{+} \quad \text{L-cysteine}^{+} \xrightarrow{k_t} Ru(bpy)_3^{2+} + \text{L-cysteine} \tag{30}$$

and the observed overall rate, k_{obs}^{calc}, is described by:

$$k_{obs}^{calc} = \frac{k_D}{1 + \frac{k_D}{K_A(k_s + k_t)}\left[\exp(\Delta G^{\ddagger}/RT) + \exp(\Delta G/RT)\right]} \tag{31}$$

where k_A is the association constant ($K_A = k_D/k_{-D}$), obtainable from:

$$K_A = \frac{4\pi N a^3}{3000} \exp(\mu(a)/kT) \tag{32}$$

with $\mu(a) = z_1 z_2 e^2/a\varepsilon_s$, where a is the distance of closest approach, ε_s is the static dielectric constant and z_1 and z_2 are the charges on the reacting species. The diffusion rate constant k_D in equation 31 has been calculated from:

$$k_D = 4\pi(D_1 + D_2)afN/1000 \tag{33}$$

where the diffusion coefficient D_i ($D_i = D_1$ or D_2) and electrostatic factor are given by:

$$D_i = kT/6\pi\eta r_i \tag{34}$$

and

$$f = \frac{\mu(a)}{kT}\left[\exp\left(\frac{\mu a}{kT} - 1\right)\right] \qquad (35)$$

where η is the viscosity of the medium and r_i is the radius of the species.

ΔG has been estimated from (36):

$$\Delta G^{\ddagger} = \frac{\Delta G}{2} + \left[\left(\frac{\Delta G}{2}\right)^2 + \Delta G^{\ddagger}(o)^2\right]^{\frac{1}{2}} \qquad (36)$$

$$\Delta G^{\ddagger}(o) = \frac{e^2}{4}\left[\frac{1}{2r_1} + \frac{1}{2r_2} - \frac{1}{a}\right]\left(\frac{1}{\varepsilon_o} - \frac{1}{\varepsilon_s}\right) \qquad (37)$$

(where $\Delta G^{\ddagger}(o)$ is the activation free energy for isoenergetic ($\Delta G = 0$) electron transfer situation and ε_o is the optical dielectric constant of the medium using the adjusted parameter $K_A(k_s + k_t) = 6.9 \times 10^8\ M^{-1}sec^{-1}$ (37).

Considering the assumptions involved the agreement between k_{obs}^{calc} and k_{obs}^{exp} (Table I) is quite remarkable.

Effects of Electrolyte Gradients on the Partitioning between the Inner and the Outer Compartments of Radicals Expelled from Vesicle Bilayers. Exit of photogenerated species from the bilayers of surfactant vesicles can be directed preferentially to the bulk solution (as opposed to the inner compartment of the vesicle) by setting up suitable potentials. Electron transfer from N-methylphenothiazine, MPTH, solubilized within the hydrophobic bilayers of DODAC surfactant vesicles, to a long chain derivative of tris(2,2'-bipyridine)ruthenium cation, $RuC_{18}(bpy)_3^{2+}$, anchored onto the inner and outer surfaces of DODAC, have been examined (21). Electron transfer resulted in the formation of N-methylphenothiazine cation radical, $MPTH^{\ddagger}$:

$$RuC_{18}(bpy)_3^{2+*} + MPTH \longrightarrow RuC_{18}(bpy)_3^{\dot{+}} + MPTH^{\dot{+}} \qquad (38)$$

The $MPTH^{\dot{+}}$ formed can disappear by a geminate type of back electron transfer at the very site of its creation:

$$RuC_{18}(bpy)_3^{\dot{+}} + MPTH^{\dot{+}} \xrightarrow{\text{geminate}} RuC_{18}(bpy)_3^{2+} + MPTH \qquad (39)$$

or due to the potential gradient exit into the DODAC entrapped water pool or into the bulk solution. The $MPTH^{\dot{+}}$ expelled into the bulk solution is long lived since electrostatic repulsion between this species and the positively charged vesicle surface decrease the probability of back reaction. Preferential expulsion of $MPTH^{\dot{+}}$

is accomplished by the addition of NaCl to the outside of already formed vesicles. This has three important consequences. First, the number of sites, where the local electrostatic field prevented the existence of $MPTH^{\dot{+}}$, is reduced. Second, a dissymmetry is created between the inner and outer surface potential of the vesicles which will increase the fraction of $MPTH^{\dot{+}}$ exiting into the bulk solution. Third, the reduced net charge on the aggregates increases the rate of back reaction. The amounts of $MPTH^{\dot{+}}$ produced and that expelled into the bulk aqueous solutions were maximized in the presence of 1.0×10^{-3} M NaCl. Under this condition there was still a sufficient electrostatic repulsion between $MPTH^{\dot{+}}$ and the charged surface of the vesicles to slow down considerably the undesirable charge recombination reactions (21).

The Role of Potential Gradients across Bilayers to Facilitate Electron Transfer. Creation of appropriate potentials also assists electron transfer across surfactant vesicle bilayers. Electron transfer from $Ru(bpy)_3^{2+*}$ to MV^{2+} (reaction 20) has been examined by placing the sensitizer on the outer surface and the electron acceptor in the inner surface (System I in Figure 3) or *vice versa* (System II in Figure 3) of anionic DHP surfactant vesicles (22). System I is much more efficient than System II. In System I all the negative charges on the inner surface of DHP vesicles are neutralized by MV^{2+}, whereas there is only partial neutralization of the outer surface by $Ru(bpy)_3^{2+}$. The gradient, created by the diffusion and Donnan potentials, facilitates the flow of electrons from the outer to the inner surface of the vesicle. The situation is quite different in System II. Vesicle-vesicle fusion precludes extensive neutralization of the outer surface by MV^{2+}, which in turn creates insufficient potential for driving the electron from the inner to the outer surface. Addition of EDTA to aqueous solution of System I resulted in the reformation of $Ru(bpy)_3^{2+}$. If PtO_2 is incorporated in the interiors of DHP vesicles of the same system, $MV^{\dot{+}}$ is reoxidized with concomitant hydrogen formation (Figure 4):

$$2MV^{\dot{+}} + H_2O \xrightarrow{PtO_2} H_2 + 2OH^- + 2MV^{2+} \tag{40}$$

Photolysis of this system leads, therefore, to the net consumption of only EDTA at very low stoichiometric $Ru(bpy)_3^{2+}$, MV^{2+} and PtO_2 concentrations (22).

Conclusion

Importance of various potential gradients in promoting charge separation in the presence of surfactant vesicles has been delineated. Semiquantitative relationships between known theories and experimental results have been demonstrated for several systems. The obtained knowledge in turn will aid the systematic optimization

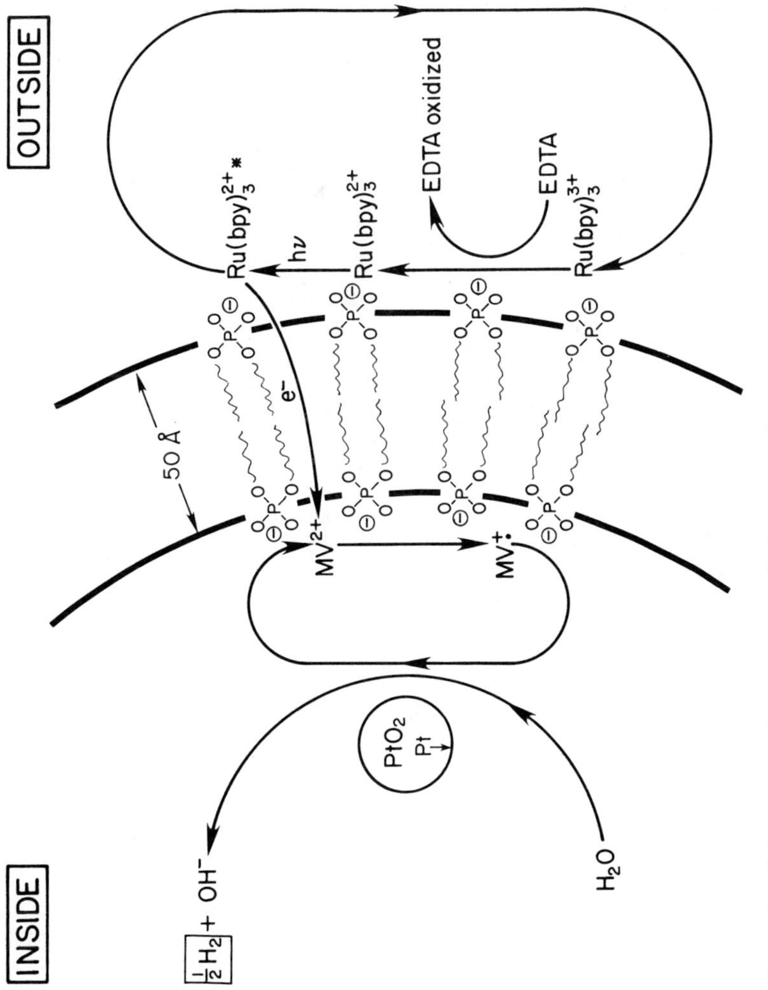

Figure 4. Schematic of the surfactant vesicle system used for the photosensitized catalytic hydrogen production, resulting only in the consumption of EDTA.

of different systems needed for viable photochemical solar energy conversions in membrane mimetic systems.

Acknowledgment

Support of this work by the Department of Energy is gratefully acknowledged.

Literature Cited

1. Porter, G.; Archer, M. V. ISR Interdiscip. Sci. Rev. 1976, 1, 119.
2. Calvin, M. Acc. Chem. Res. 1978, 11, 4701.
3. Hautala, R. R.; King, R. B; Kutal, C. "Solar Energy, Chemical Conversion and Storage"; The Humana Press: Clifton, NY, 1979.
4. Bolton, J. R. "Solar Power and Fuels"; Academic Press: New York, NY, 1977.
5. Archer, M. D. In "Photochemistry, Specialists Periodical Report"; The Chemical Society: London, 1978; Vol. 9, p. 603; *ibid*, 1977; Vol. 8, p. 570; *ibid* ; Vol. 7, p. 567; *ibid* 1976; Vol. 6, p. 736.
6. Claesson, S.; Engström, M. "Solar Energy — Photochemical Conversion and Storage"; National Swedish Board for Energy Source Development: Stockholm, 1977.
7. Barber, J. "Photosynthesis in Relation to Model Systems"; Elsevier: New York, 1979.
8. Gerischer, H.; Katz, J. J. "Light Induced Charge Separation in Biology and Chemistry"; Verlag Chemie: New York, 1979.
9. Thomas, J. K. Chem. Revs. 1980, 80, 283.
10. Grätzel, M. In "Micellization, Solubilization and Microemulsions"; (Mittal, K. L., Ed.) Plenum Press, New York, 1977; p. 531. Thomas, J. K. In "Modern Fluorescence Spectroscopy"; (Wehry, E. L., Ed.) Plenum Press: New York, 1976; p. 196. Turro, N. J.; Grätzel, M.; Braun, A. M. Angew. Chem. Int. Ed. 1980, 19, 675; Kalyanasundaram, K., Chem. Soc. Rev. 1978, 7, 435; Thomas, J. K. Acc. Chem. Res. 1977, 10, 133.
11. Willner, I.; Ford, W. E.; Otvos, J. W.; Calvin, M. Nature 1979, 280, 823.
12. Rodgers, M.A.J.; Becker, J. C. J. Phys. Chem. 1980, 84, 2762.
13. Jones, C. E.; Jones, C. A.; Mackay, R. A. J. Phys. Chem. 1979, 83, 805; Gregoritch, S. J.; Thomas, J. K. J. Phys. Chem. 1980, 84, 1491; Almgren, M.; Grieser, F.; Thomas, J. K. J. Am. Chem. Soc. 1980, 102, 3188; Kiwi, J.; Grätzel, M. J. Phys. Chem. 1980, 84, 1503; Pileni, M. P. Chem. Phys. Lett. 1980, 75, 540.
14. Kuhn, H. Pure and App. Chem. 1979, 51, 341; Mercer-Smith, J. A.; Whitten, D. G. J. Am. Chem. Soc. 1979, 101, 6620; Janzen, A. F.; Bolton, J. R. J. Am. Chem. Soc. 1979, 101, 6342; Polymeropoulos, E. E.; Möbius, D.; Kuhn, H. Thin Solid Films 1980, 68, 173.
15. Tien, H. T. "Bilayer Lipid Membranes"; Marcel Dekker: New

York, 1974; Tien, H. T., In "Topics in Photosynthesis - Photosynthesis in Relation to Model Systems"; (Barber, J., Ed.) Elsevier: Amsterdam, 1979; p. 116.
16. Meisel, D.; Matheson, M. S. J. Am. Chem. Soc. 1977, 99, 6577; Jonah, C. D.; Matheson, M. S.; Meisel, D. J. Phys. Chem. 1979, 83, 257; Meisel, D.; Rabani, J.; Meyerstein, D.; Matheson, M. S. J. Phys. Chem. 1978, 82, 985; Meisel, D.; Matheson, M. S.; Rabani, J. J. Am. Chem. Soc. 1978, 100, 117.
17. Mangel, M. Biochim. Biophys. Acta 1976, 430, 459; Toyoshima, Y.; Morino, M.; Motoki, H.; Sukigara, M. Nature 1977, 265, 187; Ford, W. E.; Otvos, J. W.; Calvin, M. Nature 1978, 274, 507; Ford, W. E.; Otvos, J. W.; Calvin, M. Proc. Natl. Acad. Sci. USA 1979, 76, 3590; Kurihara, K.; Sukigara, M.; Toyoshima, Y. Biochim. Biophys. Acta 1979, 547, 117; Kurihara, K.; Toyoshima, Y.; Sukigara, M. Biochem. Biophys. Res. Commun. 1979, 88, 320; Nagamura, T.; Takuma, K.; Tsutsui, Y.; Matsuo, T. Chem. Lett. 1980, 503; Matsuo, T.; Itoh, K.; Takuma, K.; Hashimoto, K.; Nagamura, T. Chem. Lett. 1980, 1009; Sudo, Y.; Kawashima, T.; Toda, F. J. Chem. Soc. Chem. Commun. 1979, 1044.
18. Fendler, J. H. Acc. Chem. Res. 1980, 13, 7; Herrmann, U.; Fendler, J. H. Chem. Phys. Lett. 1979, 64, 270.
19. Escabi-Perez, J. R.; Romero, A.; Lukac, S.; Fendler, J. H. J. Am. Chem. Soc. 1979, 101, 2231.
20. Nomura, T.; Escabi-Perez, J. R.; Sunamoto, J.; Fendler, J. H. J. Am. Chem. Soc. 1980, 102, 1484.
21. Infelta, P. P.; Grätzel, M.; Fendler, J. H. J. Am. Chem. Soc. 1980, 102, 1479.
22. Monserrat, K., Grätzel, M. and Tundo, P. J. Am. Chem. Soc. 1980, 102, 5527.
23. Tunuli, M. S.; Fendler, J. H. J. Am. Chem. Soc. 1981, 103, 000.
24. Pileni, M.-P. Chem. Phys. Lett. 1980, 71, 317.
25. Takayanagi, T.; Nagamura, T.; Matsuo, T. Chem. Lett. 1980, 503.
26. Monserrat, K.; Grätzel, M. J. Chem. Soc. Chem. Commun. 1981, 183.
27. Nagamura, T.; Matsuo, T.; Fendler, J. H. Chem. Phys. Lett. 1981, in press.
28. Fendler, J. H. J. Phys. Chem., 1980, 84, 1485.
29. Newman, J. S. "Electrochemical Systems"; Prentice-Hall, Inc.: Englewood Cliffs, NJ, 1973.
30. Goldman, D. E. J. Gen. Physiol. 1943, 27, 37.
31. Mott, N. F. Proc. Roy. Soc. London 1939, A171, 27.
32. Schlögl, R.; Helfferich, F. Z. Electrochem. 1952, 56, 644; Schlögl, R. Z. Physical Chem. N. F. 1954, 1, 305.
33. Porter, G. Proc. Roy. Soc. London, Ser. A. 1978, 362, 281; Porter, G. Pure Appl. Chem. 1978, 50, 263.
34. Adam, G.; Delbrück, M. In "Structural Chemistry and Molecular Biology"; (Rich, A.; Davidson, N., Eds.) Freeman and Co.: San Francisco, CA, 1968; Richter, P. H.; Eigen, M. Biophys. Chem. 1974, 2, 225; Eigen, M. in "Quantum Statistical Mechanics in the Natural Sciences"; (Kuseneglu, B.; Mintz,

S. C.; Widmayer, S., Eds.); Plenum Press: New York, 1974; p. 37.
35. Tunuli, M. S.; Fendler, J. H. unpublished results, 1981.
36. Marcus, R. A. J. Chem. Phys. 1965, 43, 679.
37. Rehm, D.; Weller, A. Israel J. Chem. 1970, 8, 259.

RECEIVED September 9, 1981.

5

Control of Photosensitized Electron Transfer Reactions in Organized Interfacial Systems

Vesicles, Water-in-Oil Microemulsions, and Colloidal Silicon Dioxide Particles

ITAMAR WILLNER, COLJA LAANE, JOHN W. OTVOS, and MELVIN CALVIN

University of California, Laboratory of Chemical Biodynamics, Department of Chemistry and Lawrence Berkeley Laboratory, Berkeley, CA 94720

The separation of photoproducts formed in photosensitized electron transfer reactions is essential for efficient energy conversion and storage. The organization of the components involved in the photoinduced process in interfacial systems leads to efficient compartmentalization of the products. Several interfacial systems, e.g., lipid bilayer membranes (vesicles), water-in-oil microemulsions and a solid SiO_2 colloidal interface, have been designed to accomplish this goal.

An electron transfer across a lipid bilayer membrane leading to the separation of the photoproducts at opposite sides of the membrane is facilitated by establishing a transmembrane potential and organizing the cotransport of cations with specific carriers.

In the water-in-oil microemulsion the separation of photoproducts is achieved by means of the hydrophilic-hydrophobic nature of the products. A two compartment model system to accomplish the photodecomposition of water is described. Photosensitized electron transfer reactions analogous to those occurring in the two half-cells are presented. In these systems the phase transfer of one of the photoproducts into the continuous oil phase is essential to stabilize the photoproducts.

The colloidal SiO_2 particles provide a charged interface that interacts with charged photoproducts. By designing a system that results in oppositely charged photoproducts, we can produce a retardation of recombination by the charged interface. The photosensitized reduction of a neutral acceptor, propylviologen sulfonate (PVS^o) by positively charged sensitizers such as $Ru(bipy)_3^{2+}$ and Zn-tetramethylpyridinium porphyrin, $Zn-TMPyP^{4+}$, is des-

0097-6156/82/0177-0071$06.00/0

cribed. The reactions are substantially enhanced in the SiO_2 colloid as compared to those in the homogeneous phase. The effect of the SiO_2 interface is attributed to a high surface potential that results in the separation of the intermediate photoproducts. The quantum yields of the photosensitized reactions are correlated to the interfacial surface potential and the electrical effects of other charged interfaces such as micelles are compared with those of SiO_2.

The possible utilization of the energy stored in the stabilized photoproducts in further chemical reactions is discussed. Special attention is given to the photodecomposition of water as a reaction route.

Reactions in organized media as a means of modeling natural processes are currently an intensive subject of research (1,2). Of particular interest is the subject of "Artificial Photosynthesis", that is, an attempt to create a synthetic apparatus that mimics the functions of the natural process (3,4). The natural photosynthetic cycle leading to the production of carbohydrates (eq.1) can be separated into two major parts (5): a photochemical part, in which visible light is captured and transformed into chemical energy, and a chemical part in which the stored energy is utilized in a sequence of dark reactions. The process is summarized in the "Z-scheme" (Figure 1). The chloroplast utilizes two photosystems composed of well organized pigment molecules. Photoexcitation of these units induces electron transfer reactions that result in the oxidation of water to oxygen and formation of a reduced intermediate (ferredoxin). The reducing power is then used in a set of dark reactions to form carbohydrates from CO_2.

$$H_2O + CO_2 \xrightarrow{h\nu} [CH_2O] + O_2 \qquad \text{(eq.1)}$$

In this article we will discuss several approaches to the design of organized and controlled photosensitized electron transfer reactions. Special emphasis will be given to the utilization of the stored energy in the photodecomposition of water (eq.2).

$$H_2O \xrightarrow{h\nu} H_2 + 1/2\ O_2 \qquad \text{(eq.2)}$$

A schematic cycle describing the principle of light capture and energy storage via a photosensitized electron transfer process in an artificial device is presented in Figure 2. In this system a synthetic sensitizer, S, substitutes for the natural chlorophyll as the light capturing entity. Excitation of the sensitizer, followed by an electron transfer to electron acceptor, A, results in the oxidized sensitizer and a reduced species, A^-. Oxidation of an electron donor, D, recycles the sensitizer and produces an

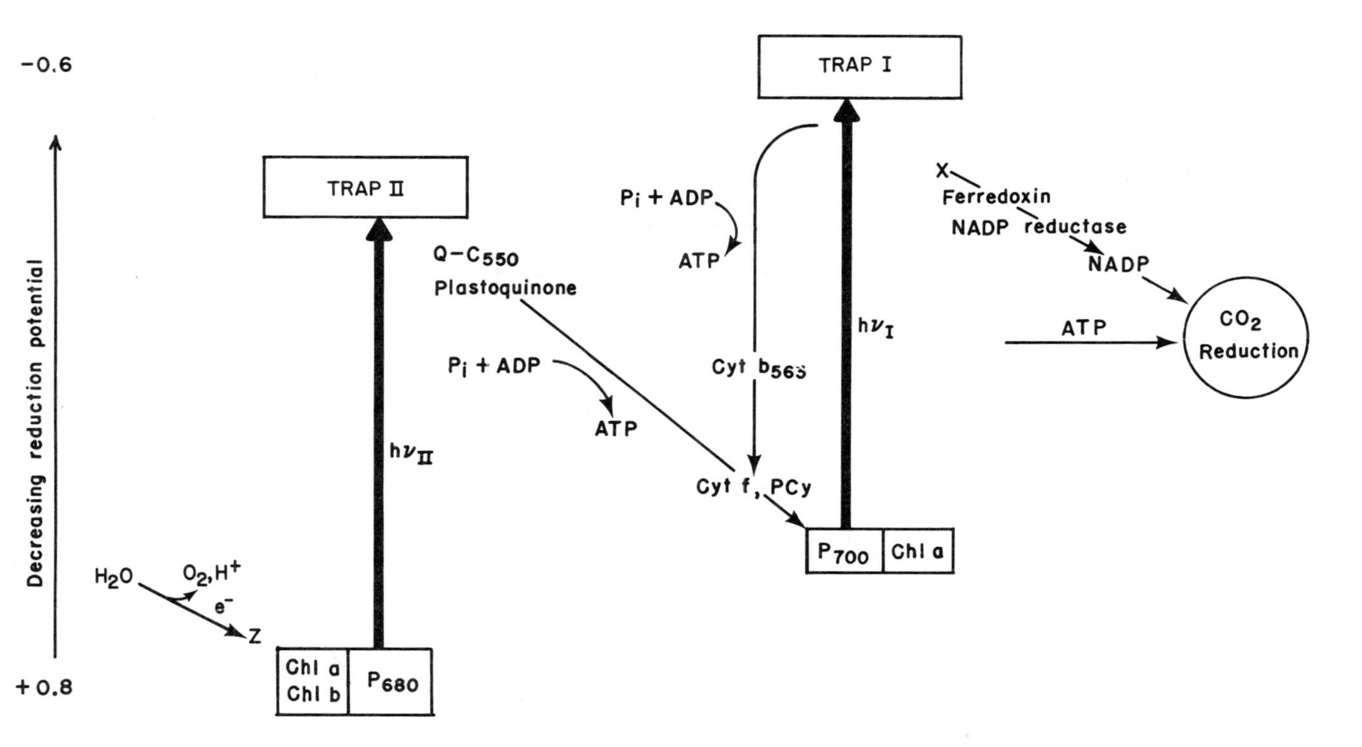

Figure 1. Electron transfer (Z-scheme) in photosynthesis.

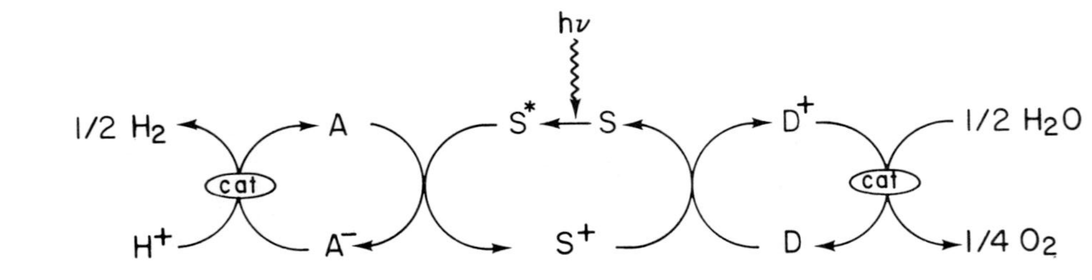

Figure 2. Cyclic photochemical scheme for water decomposition where S represents an artificial sensitizer that simulates the function of the natural chlorophyll and Cat represents a catalyst.

oxidized product, D^+. In such a way the light energy is transformed into chemical energy and stored in the reduced and oxidized products. The reduced electron acceptor and oxidized electron donor can then be further used to produce hydrogen and oxygen from water, as one example of long term energy storage. In this way all the active components of the system are recycled and the net result is the conversion of water to a potential fuel (hydrogen). However, a basic limitation of such a system is the thermodynamically favored back-electron transfer reactions of the intermediate photoproducts (eq.3 and 4). By these pathways the energy stored in the photochemical event is degraded and an efficient utilization of the photoproducts is prevented.

$$A^- + S^+ \longrightarrow A + S \qquad \text{(eq.3)}$$
$$A^- + D^+ \longrightarrow A + D \qquad \text{(eq.4)}$$

In the natural process this limitation is solved by an organization of the components in a membrane. The two photosystems function at opposite sides of a membrane and consequently the intermediates are separated and an efficient chemical utilization is feasible.

Thus, mimicking photosynthesis with the goal of decomposing water must involve three cooperative elements: (a) A light capturing entity that is capable of photosensitizing electron transfer reactions; (b) an interfacial barrier that separates the intermediate photoproducts and prevents their recombination; and (c) suitable redox catalysts capable of reducing and oxidizing water.

Many synthetic dyes such as porphyrins, acridine, thionine and flavin dyes have been used in photosensitization of electron transfer reactions. In the past few years several promising organometallic compounds have been prepared as substitutes for the natural labile chlorophyll. These organometallics include a variety of metals, chelated to bipyridine or porphyrin ligands (6). The photophysical properties of these sensitizers and their potential use in artificial photosynthetic devices have been extensively reviewed (7, 8, 9). In particular, sensitizers such as Ru(II)-tris-bipyridine, $Ru(bipy)_3^{2+}$(1), and Zn-porphyrins, such as Zn-meso-tetraphenylporphyrins (2) or water-soluble derivatives (3) and (4) have been widely explored. Different structural modifications such as hydrophobic substituents and charged headgroups have also been introduced. Thus, control of electrostatic interactions and precise location of the sensitizer in hydrophilic or hydrophobic environments can be achieved.

The storage of energy by means of a photosensitized electron transfer cycle as presented in Figure 2 requires a close proximity of the components for efficient quenching of the excited species. However, once the photoproducts are produced, their separation must be assisted and a barrier for their recombination must be introduced. Several interfacial systems such as micelles (10,11), water-in-oil (12) or oil-in-water microemulsions (13) and bilayer

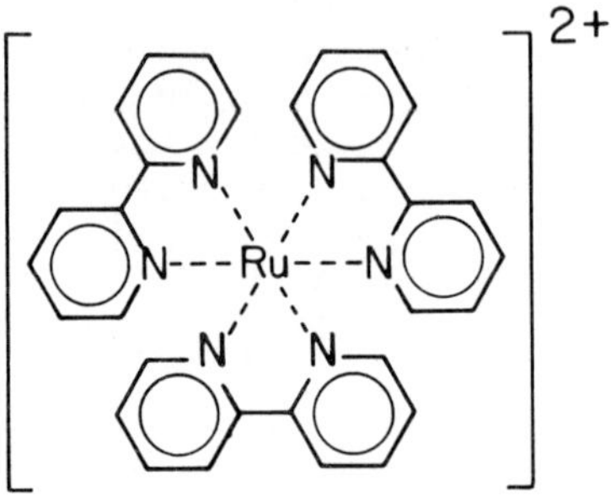

2+

N N N Ru N N N

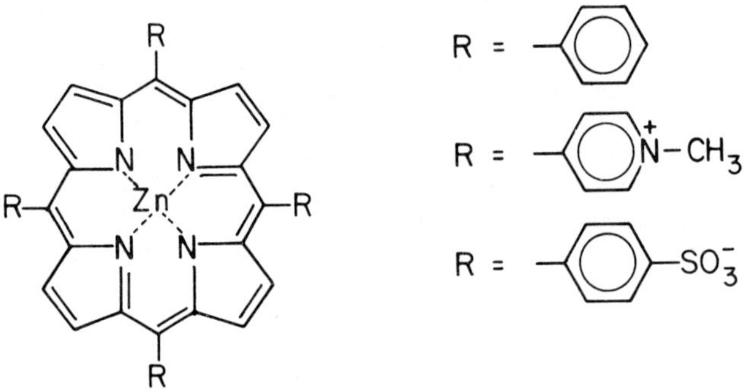

(1)

R

N N Zn N N

R R

R

R = (2)

R = $\overset{+}{N}-CH_3$ (3)

R = $-SO_3^-$ (4)

$\overset{+}{N}$ $\overset{+}{N}$

SO_3^- ^-O_3S

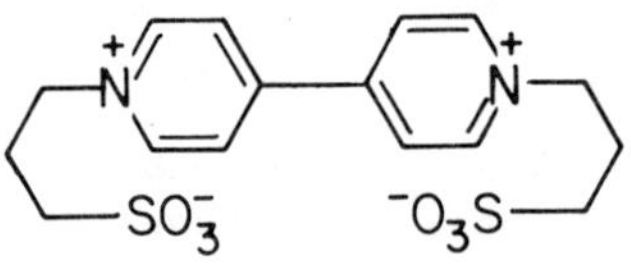

(5)

membranes (14,15) (vesicles) provide microenvironments that meet these requirements.

Since these interfaces are usually constructed of charged detergents a diffuse electrical double layer is produced and the interfacial boundary can be characterized by a surface potential. Consequently, electrostatic as well as hydrophilic and hydrophobic interactions of the interfacial system can be designed. In this report we will review our achievements in organizing photosensitized electron transfer reactions in different microenvironments such as bilayer membranes and water-in-oil microemulsions. In addition, a novel solid-liquid interface, provided by colloidal SiO_2 particles in an aqueous medium will be discussed as a means of controlling photosensitized electron transfer reactions.

Photosensitized Electron Transfer Across Bilayer Membranes

With the knowledge that membranes play an important role in the natural process, we initiated a study in which bilayer phospholipid membranes (vesicles) serve as an artificial structure. For this purpose an electron transfer across the bilayer boundary must be accomplished (14). The schematic of our system is presented in Figure 3. In this system an amphiphilic Ru-complex is incorporated into the membrane wall. An electron donor, EDTA, is entrapped in the inner compartment of the vesicle, and heptylviologen (HV^{2+}) as electron acceptor is introduced into the outer phase. Upon illumination an electron transfer process across the vesicle walls is initiated and the reduced acceptor ($HV^{\dagger}$) is produced. The different steps involved in this overall reaction are presented in Figure 3. The excited sensitizer transfers an electron to HV^{2+} in the primary event. The oxidized sensitizer thus produced oxidizes a Ru^{2+} located at the inner surface of the vesicle and thereby the separation of the intermediate photoproducts is assisted (14). The further oxidation of EDTA regenerates the sensitizer and consequently the separation of the reduced species, $HV^{\dagger}$, from the oxidized product is achieved. In this system the basic principle of a vectorial electron transfer across a membrane is demonstrated. However, the quantum yield for the reaction is rather low ($\emptyset = \sim 4 \times 10^{-4}$).

The transmembrane electron transfer was found to be the rate limiting factor for the overall reaction and the origin of the low efficiency. The electron transfer across the membrane must be followed by cotransport of cations in order to keep charge neutrality. Since the membrane has a low permeability to such cations, the photosensitized reaction might be limited by this effect. Indeed, further elaboration of the vesicle system by including cation carriers can improve the photoinduced reaction. For this purpose hydrophobic cation carriers (ionophores) such as valinomycin (specific for K^+), CCCP (specific for H^+) and gramicidin (transport agent of K^+, Na^+ and H^+) have been incorporated into the hydrophobic region of the vesicles (16). The photosensitized electron

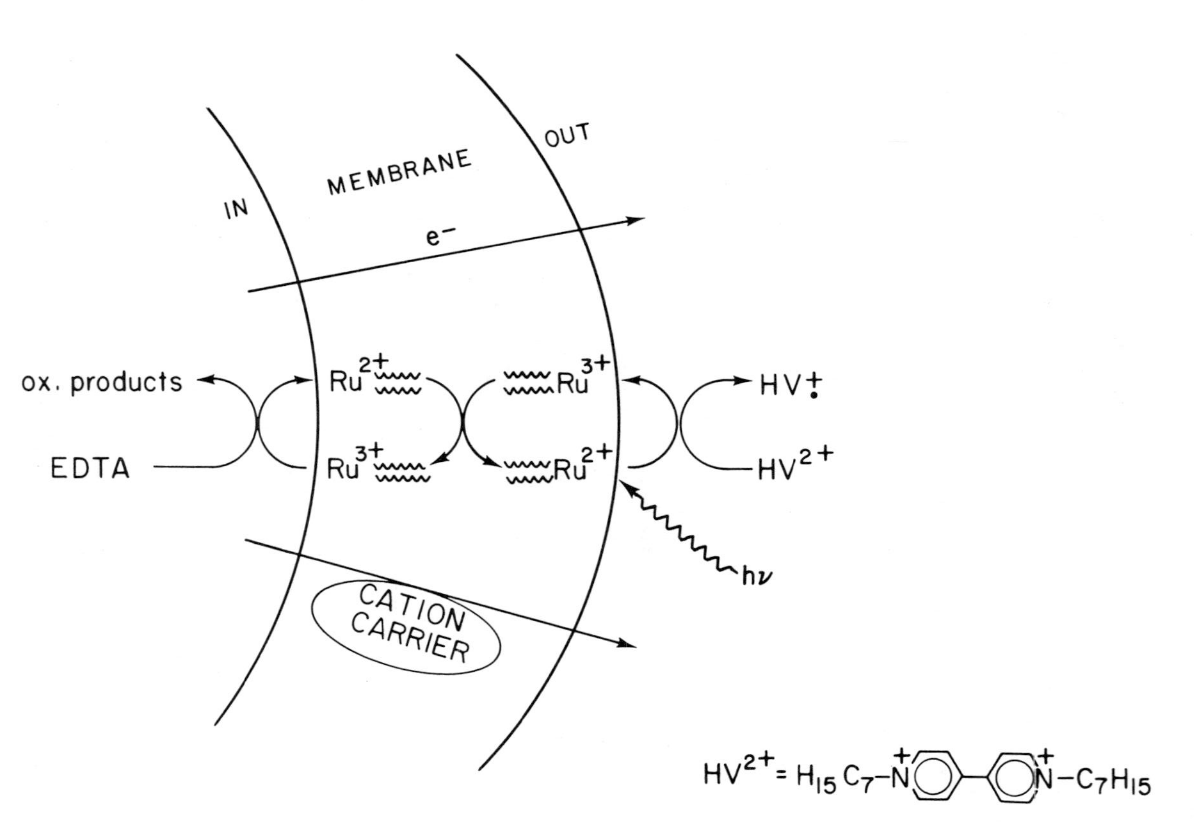

Figure 3. Scheme for photosensitized electron transfer across a lipid vesicle wall.

transfer reaction in the presence of these carriers is enhanced to a 3-6 fold extent, depending on the ionophore (Figure 4A). These results confirm that the cotransport of cations plays an important role in the photosensitized reactions.

In addition to their function in establishing charge neutralization during the photochemical reactions, cations might, by proper organization, even assist the electron transfer via production of a transmembrane potential. Different concentrations of K^+ in the opposite aqueous phases of the lipid bilayer were used to test for this effect. The specific K^+ carrier, valinomycin, was incorporated into the vesicle walls. Consequently, owing to the concentration difference, a long-lasting transmembrane potential is established. The photosensitized electron transfer reaction appears to be affected by such electric fields (Figure 4B). It can be seen that when the ratio of $K^+_{in}/K^+_{out} > 1$, i.e., interior boundary negative relative to the exterior, the reaction is 2-fold enhanced as compared to the system without any applied field. Conversely, when the vesicles are designed such that $K^+_{in}/K^+_{out} < 1$ and an opposite potential is formed, the quantum yield is decreased and approaches the value obtained in the absence of any valinomycin (16). We can see that the combined effects of cation permeability and a transmembrane potential result in an 11-fold enhancement in the photosensitized reduction of heptylviologen (HV^{2+}). Thus, proper organization of different components in the lipid bilayer interfacial system can enhance electron transfer reactions and assist the separation of photoproducts across the bilayer.

Photosensitized Electron Transfer in Water-in-Oil Microemulsions

A water-in-oil microemulsion is an interfacial system of aqueous droplets in a continuous oil phase. By including water soluble or amphiphilic reagents in this system, the components can be concentrated at will in the different phases of the microemulsion. By selecting an electron donor or electron acceptor that alters its amphiphilic properties upon oxidation or reduction, one of the photoproducts can be extracted into the continuous organic phase and so the separation of the photoproducts in two distinct phases is achieved.

A possible model system is composed of two compartment that include water-in-oil microemulsions represented in Figure 5 as two droplets. In the aqueous phases of the two compartments two different sensitizers, S_1 and S_2 are solubilized. In one compartment, an electron acceptor, A_2, is solubilized in the aqueous phase while the electron donor, D_2, is concentrated at the interface of the microemulsion. In the complementary half-cell the electron donor, D_1, is solubilized in the water droplets and the electron acceptor, A_1, is localized at the interface. The electron donor, D_2, and electron acceptor, A_1, are designed in such a way that oxidized D_2 and reduced A_1 are extracted into the continuous organic phase. The photosensitized reactions initiated in the two

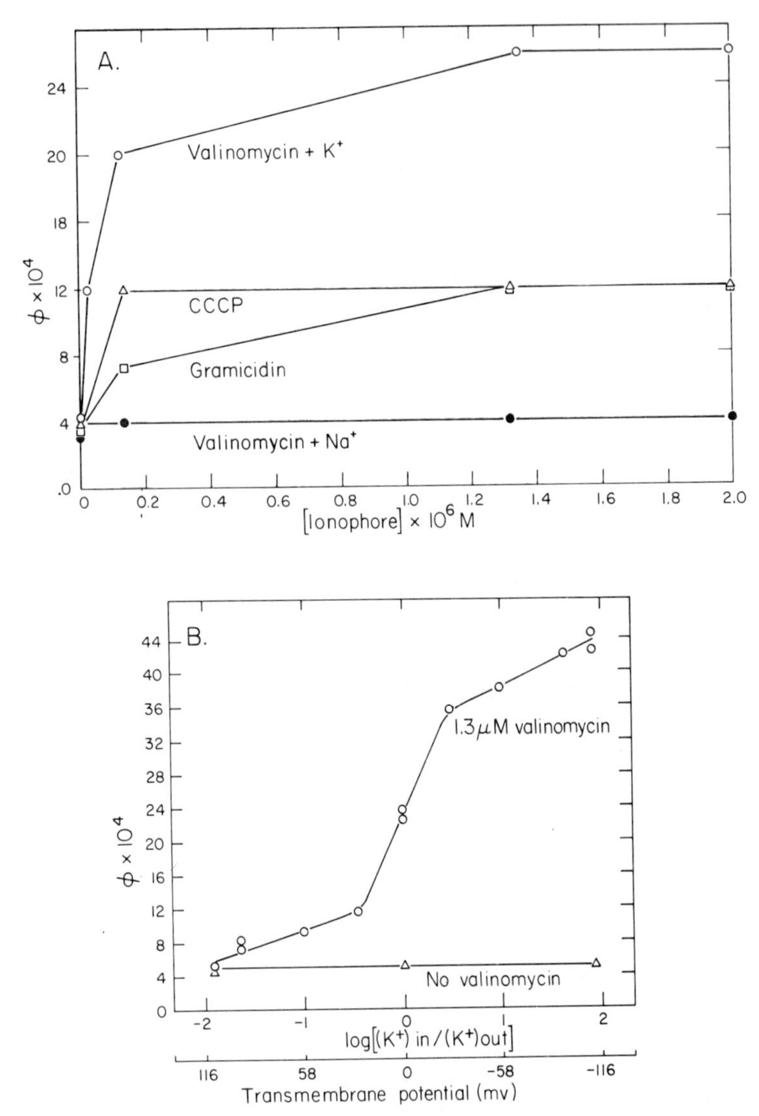

Figure 4. Effect of ionophores (A) and transmembrane potentials (B) on quantum yield of heptylviologen reduction in the vesicle system where φ is the quantum yield, CCCP (carbonyl cyanide m*-chlorophenylhydrazone) is the H^+ carrier, and valinomycin is the K^+ carrier; gramicidin makes the membrane permeable for cations such as H^+, K^+, and Na^+.*

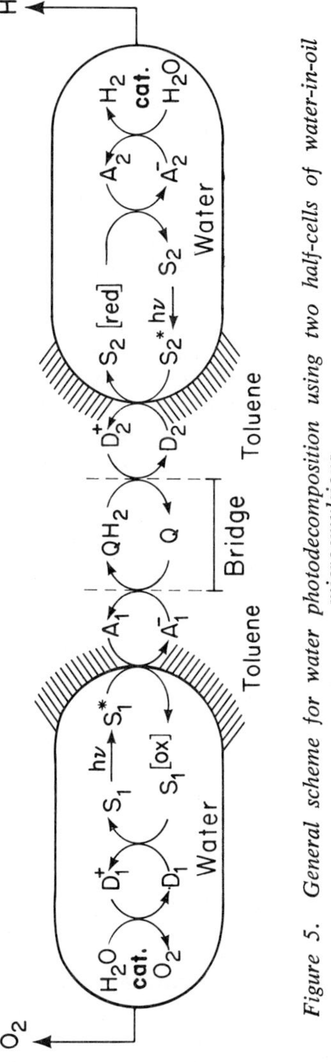

Figure 5. General scheme for water photodecomposition using two half-cells of water-in-oil microemulsions.

half-cells thus result in the separation of the photoproducts. Utilization of the water soluble photoproducts in decomposing water and coupling the two compartments with an electron mediator and proton carrier should recycle all the ingredients of the system except water.

The two half-cells proposed in this system have been constructed separately. The analogue for the oxidative half-cell (13) (Figure 6A) was composed of the electron donor EDTA and the sensitizer $Ru(bipy)_3^{2+}$ (1), soluble in the water droplet. As electron acceptor, benzylnicotinamide, BNA^+, was used. This amphiphilic compound is expected to concentrate at the water-oil interface. The photosensitized electron transfer reaction forms the reduced lipophilic electron acceptor BNA· which is ejected into the continuous organic phase and thus separated from the oxidized product. In order to monitor the entire phase transfer of the reduced acceptor, BNA·, a secondary electron acceptor, p-dimethylaminoazobenzene (dye), was solubilized in the continuous oil phase. The photochemically induced electron transfer reaction in this system results in the reduction of the dye ($\phi = 1.3 \times 10^{-3}$). Exclusion of the sensitizer or EDTA or the primary electron acceptor, BNA^+, from the system resulted in no detectable reaction. Substitution of the primary acceptor with a water soluble derivative, N-propylsulfonate nicotinamide, similarly results in no reduction of the dye. These results indicate that to accomplish the cycle formulated in Figure 6A the amphiphilic nature of the primary electron acceptor and its phase transfer ability in the reduced form are necessary requirements.

Similarly, the reduction half-cell of the model has been constructed. In the latter system (Figure 6B) the electron acceptor dimethyl-4,4'-bipyridinium (methylviologen, MV^{2+}) and $Ru(bipy)_3^{2+}$ (1) or the water soluble Zn-porphyrins (3) and (4) were dissolved in the aqueous phase of the water-in-oil microemulsion with the electron donor, thiophenol, being concentrated at the water-oil interface. Illumination of this system results in the production of the viologen radical cation. This photosensitized electron transfer process results in the separation of the reduced photoproduct, $MV^{\dot{+}}$, from the oxidized product, diphenyldisulfide, which is in the toluene phase.

Photosensitized Electron Transfer Reactions in SiO_2 Colloids

Colloidal SiO_2 particles in an aqueous suspension provide a solid-liquid interface. The silanol groups on the particle surface are ionized at a pH $\geqslant 6$. Consequently, the surface of the particle is negatively charged and a diffuse electrical double layer is produced in the vicinity of the solid interface (17,18). Because of the negative charges on the particles they repel one another and their agglomeration is prevented. The particles can be used to exert electrostatic repulsive and attractive interactions with the components involved in photosensitized reactions.

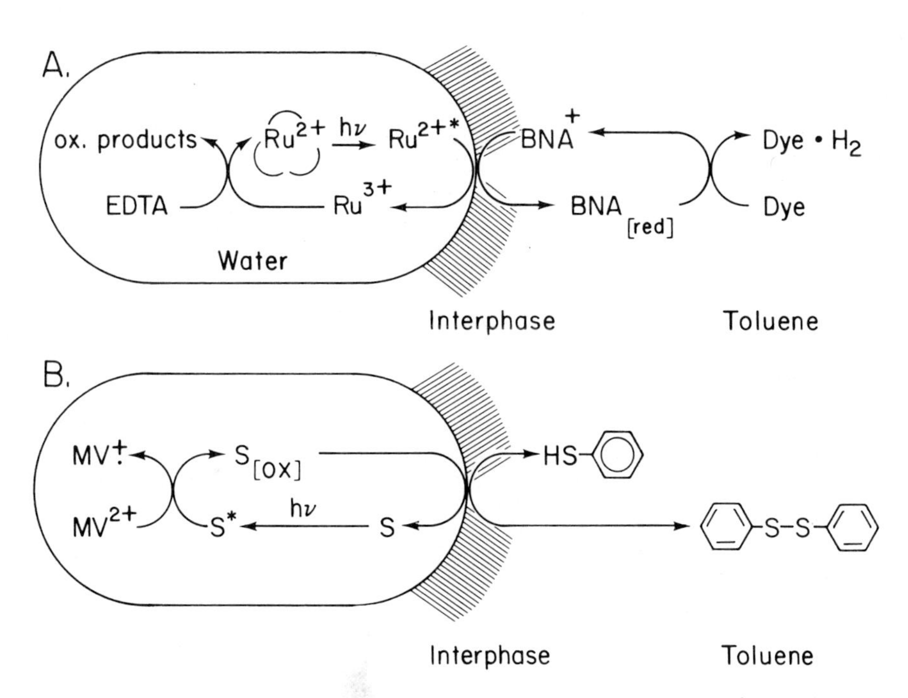

Figure 6. Cyclic mechanism for photoinduced electron transfer across the interface of a water-in-toluene microemulsion. Key: A, oxidative half-cell; and B, reductive half-cell.

By means of these interactions a component can be selectively adsorbed to the interface and its recombination with an oppositely charged photoproduct can be retarded.

To achieve such an organization in the system the different components have to be functionalized. Two positively charged sensitizers, $Ru(bipy)_3^{2+}$ (1) or Zn-meso-tetramethylpyridinium porphyrin, $Zn\text{-}TMPyP^{4+}$ (3), that are adsorbed to the SiO_2 interface are used (19). The zwitterionic dipropylsulfonate-4,4'-bipyridinium (5) (propylviologen sulfonate, PVS^o) is used as electron acceptor, and triethanolamine, TEA, is introduced as electron donor. Photosensitized electron transfer in these systems results in a rapid production of the viologen radical, $PVS^{\overline{\cdot}}$. The rates of $PVS^{\overline{\cdot}}$ formation in the colloidal SiO_2 systems using the different sensitizers are shown in Figure 7, and compared with the analogous reactions in a homogeneous phase. It can be seen that the electron transfer reactions in the SiO_2 colloid are ca. 10-fold enhanced relative to the homogeneous phase and using $Zn\text{-}TMPyP^{4+}$ as sensitizer a high quantum yield (ϕ = 0.35) is obtained. The enhanced quantum yields in the SiO_2 colloids are ascribed to the control of the electron transfer reaction by means of electrostatic interactions (eq. 5 and 6 and Figure 8). The electron transfer from the excited sensitizer, $Ru(bipy)_3^{2+}$, to the neutral electron acceptor results in two oppositely charged photoproducts. The charged interface interacts with these intermediate photoproducts; the oxidized sensitizer is adsorbed at the interface while the reduced, negatively charged electron acceptor is repelled. Consequently, the electrostatic interactions introduce a barrier to the degradative geminate recombination of the photoproducts. As a result, the further effective utilization of the oxidized product, $Ru(bipy)_3^{3+}$, in oxidizing the electron donor, TEA, is facilitated and high quantum yields are obtained.

$$Zn\text{-}TMPyP^{4+} + PVS^{o} \underset{k_b}{\overset{h\nu}{\rightleftarrows}} Zn\text{-}TMPyP^{5+} + PVS^{\overline{\cdot}} \quad \text{(eq.5)}$$

$$Ru(bipy)_3^{2+} + PVS^{o} \underset{k_b}{\overset{h\nu}{\rightleftarrows}} Ru(bipy)_3^{3+} + PVS^{\overline{\cdot}} \quad \text{(eq.6)}$$

The function of the SiO_2 colloid in retarding back-electron transfer reactions has been confirmed by several methods:

(a) The quantum yield in the SiO_2 colloid depends strongly on the ionic strength of the medium. By increasing the ionic strength the interfacial surface potential is decreased. As a result, the electrostatic interactions with the interface are reduced and the quantum yield is decreased.

(b) Substitution of the positive sensitizer with one that is negatively charged yields two photoproducts that are repelled by the interface. Thus, the function of the interface in separating the active intermediates is lost. Indeed, with a negatively charged sensitizer, Zn-meso-tetraphenylporphyrin sulfonate, $Zn\text{-}TPPS^{4-}$ (4),

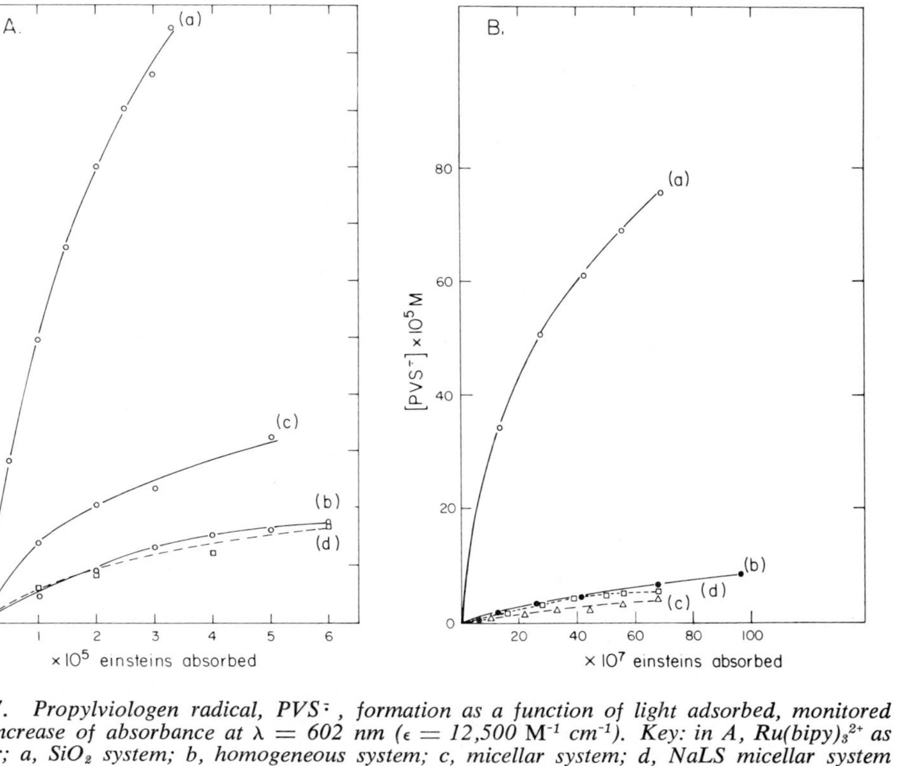

Figure 7. Propylviologen radical, $PVS^{\cdot-}$, formation as a function of light adsorbed, monitored by the increase of absorbance at $\lambda = 602$ nm ($\epsilon = 12{,}500$ M^{-1} cm^{-1}). Key: in A, $Ru(bipy)_3^{2+}$ as sensitizer; a, SiO_2 system; b, homogeneous system; c, micellar system; d, NaLS micellar system with 0.1 M NaCl; and in B, Zn-$TMPyP^{4+}$ and Zn-$TPPS^{4-}$ as sensitizers; a, SiO_2 system with Zn-$TMPyP^{4+}$; b, homogeneous system with Zn-$TMPyP^{4+}$; c, SiO_2 system with Zn-$TPPS^{4-}$; d, homogeneous system with Zn-$TPPS^{4-}$.

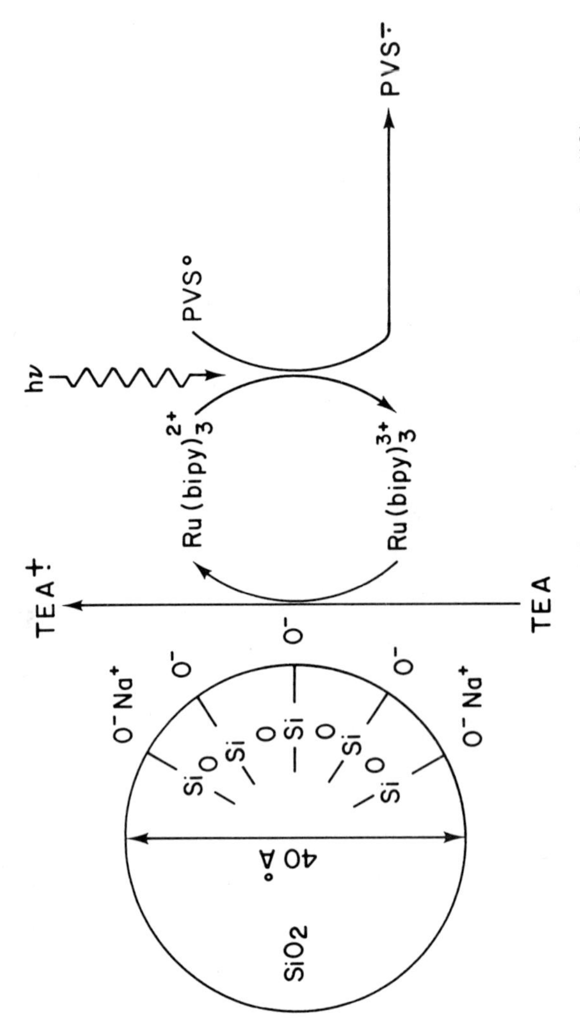

Figure 8. Schematic function of SiO_2 particles in separating photoproducts (19).

there is no enhancement of quantum yield in the SiO_2 system (Figure 7B).
(c) The back-electron transfer reaction of the intermediate photoproducts (eq. 5 and 6) has been directly followed in the SiO_2 colloid by means of flash photolysis and compared to the similar process in a homogeneous phase. A significant retardation of back-electron transfer is observed. With Zn-$TMPyP^{4+}$ as sensitizer, the recombination rate constant (eq.5) in the SiO_2 colloid is reduced by a factor of 100 relative to the value in a homogeneous phase. Similarly, with $Ru(bipy)_3^{2+}$ as sensitizer the recombination rate is ca. 90-fold retarded in the SiO_2 colloid.

The extent to which back-electron transfer reactions are retarded in the SiO_2 colloid can be improved by introduction of multinegatively charged electron acceptors such as $Fe(CN)_6^{3-}$ that increase the repulsive interactions with the interface (20). However, with such electron acceptors the primary electron transfer event is expected to be rather inefficient because they cannot approach the interface. In order to keep the balance of efficient quenching of the excited state, together with a substantial retardation of the recombination rate, two coupled electron acceptors can be used. For this purpose, a colloidal SiO_2 system has been designed in which $Ru(bipy)_3^{2+}$ is the sensitizer, PVS^o (5) the primary electron acceptor and triethanolamine, TEA, the electron donor. A secondary electron acceptor, $K_3Fe(CN)_6$, is introduced into the system to provide a sink for the electron (Figure 9). The complete photosensitized electron transfer process results in the reduction of $Fe(CN)_6^{3-}$ to $Fe(CN)_6^{4-}$ (Figure 10). It appears that the photosensitized reaction is at least 60-fold enhanced relative to the reaction in a homogeneous phase (20). The sequence of events occurring in this photosensitized electron transfer have been followed by flash photolysis. The reduced primary electron acceptor $PVS^{\overline{\cdot}}$ is produced by the quenching of the excited sensitizer adsorbed to the SiO_2 interface (eq.6). The reduced species is ejected into the continuous aqueous phase where $Fe(CN)_6^{3-}$ is reduced (eq. 7) in a "dark" reaction. The intermediate photoproducts thus created, $Ru(bipy)_3^{3+}$ and $Fe(CN)_6^{4-}$, tend to back-react (eq.8). In a homogeneous system this process is diffusion controlled ($k_b \approx 10^{10} M^{-1} \cdot sec^{-1}$)(21). However, in the SiO_2 colloid a substantial inhibition of the recombination rate is observed ($k_b = 10^6$-10^7 $M^{-1} \cdot sec^{-1}$). These results indicate that the different functions required for an efficient electron transfer process can be achieved by coupling two or more electron acceptors.

$$PVS^{\overline{\cdot}} + Fe(CN)_6^{3-} \longrightarrow PVS^o + Fe(CN)_6^{4-} \quad \text{(eq.7)}$$

$$Ru(bipy)_3^{3+} + Fe(CN)_6^{4-} \xrightarrow{k_b} Ru(bipy)_3^{2+} + Fe(CN)_6^{3-} \quad \text{(eq.8)}$$

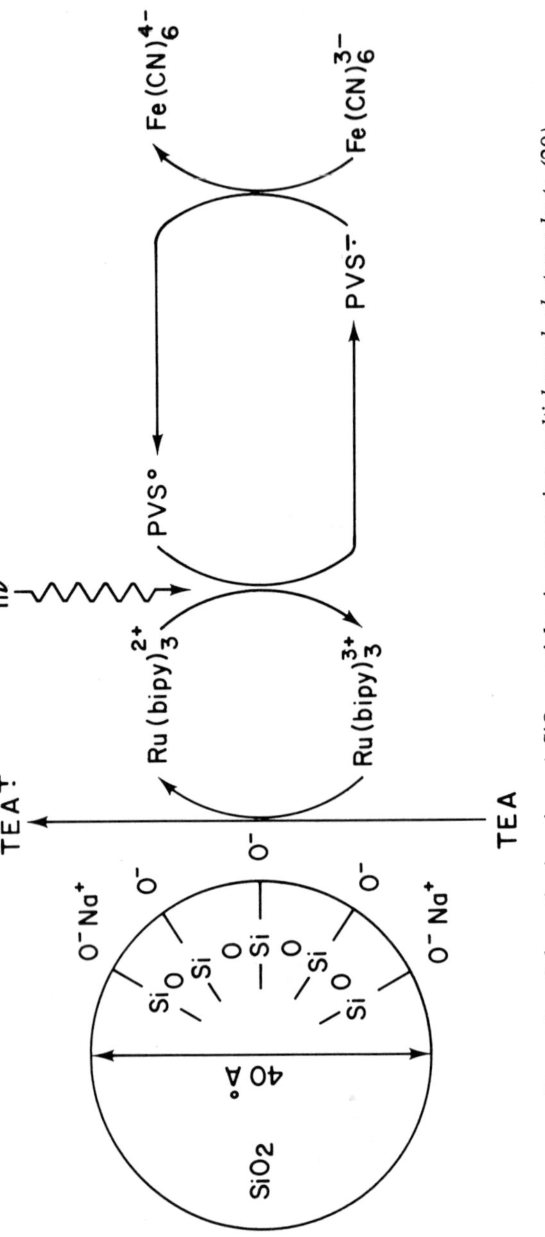

Figure 9. Schematic function of SiO_2 *particles in separating multicharged photoproducts (20).*

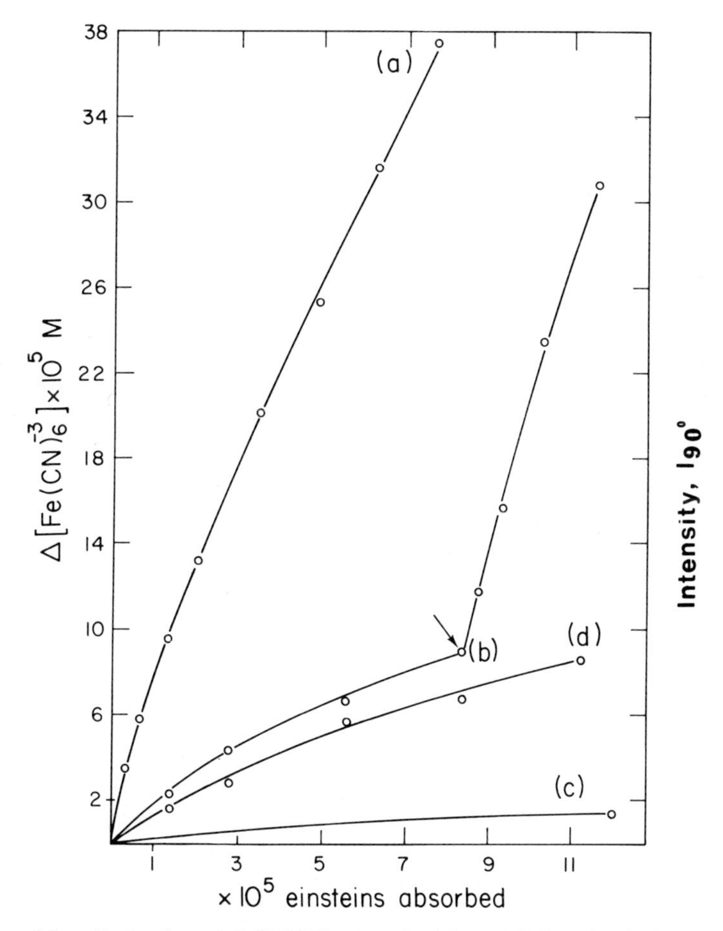

Figure 10. Reduction of $K_3Fe(CN)_6$ as a function of light adsorbed. Key: a, SiO_2 system including PVS°; b, SiO_2 system; arrow, time of PVS° addition; c, homogeneous system; and d, NaLS micellar system (20).

Correlation of Quantum Yields with Interfacial Potentials

The function of the SiO_2 colloid in the photosensitized electron transfer originates from selective interactions of the components with the interface. The electrical properties of the interface and the binding characteristics of the positively charged sensitizer, $Ru(bipy)_3^{2+}$, have been examined by means of flow dialysis ($K_{ass.} = 1.1 \times 10^2\ M^{-1}$)(22). The number of binding sites on each SiO_2 particle has been determined to be 65. These ionic sites establish an interfacial surface potential of ca. -170 mV.

The quantum yield for the photosensitized reduction of PVS^o (using $Ru(bipy)_3^{2+}$ as sensitizer) has been correlated with the interfacial surface potential of the SiO_2 colloid (controlled by varying the ionic strength of the medium) (22). The correlation curve (Figure 11) shows that up to an interfacial potential of ca. -40 mV the quantum yield is not affected. Increasing the potential above this apparent threshold value results in a sharp increase in the quantum yield. A similar correlation curve was obtained when $Zn\text{-}TMPyP^{4+}$ was used as sensitizer instead of $Ru(bipy)_3^{2+}$.

The organization of components in the SiO_2 colloids and the electrostatic interactions could, in principle, be designed with other negatively charged interfaces such as micelles. The photosensitized reduction of PVS^o using $Ru(bipy)_3^{2+}$ as sensitizer and triethanolamine, TEA, as electron donor has been investigated in the presence of negatively charged NaLS micelles and compared to the results in the SiO_2 colloid (Figure 7A) (22). The size of the NaLS micelles is similar to that of the SiO_2 particles. The sensitizer, $Ru(bipy)_3^{2+}$, appears to bind firmly to the micellar interface ($K_{ass.} = 3.5 \times 10^3\ M^{-1}$). Yet,the quantum yield for the $PVS^{\bar{\cdot}}$ formation is 4-fold less efficient than that observed in the SiO_2 colloid. This result is attributed to the difference in the surface potential of the two interfaces. Flow dialysis measurements (22) indicate that the NaLS micellar interface has a surface potential of only -85 mV, significantly lower than the value determined for the SiO_2 interface (-170 mV). The experimental quantum yields in the NaLS micellar system fit nicely into the correlation curve shown in Figure 11. This indicates that due to the relatively low surface potential of the micelles the electrostatic interactions are not as effective. The comparison of photoinduced reactions in the SiO_2 colloid to that occurring in the NaLS micelles implies that both interfaces are capable of exerting electrostatic interactions. This can be used for organizing the components involved in the photochemical reaction. However, the physical characteristics of the electric field of the different interfaces is rather important in controlling the reaction. In the NaLS micellar system, despite the organization of the components, the surface potential is relatively low and limits the ability to retard back reaction.

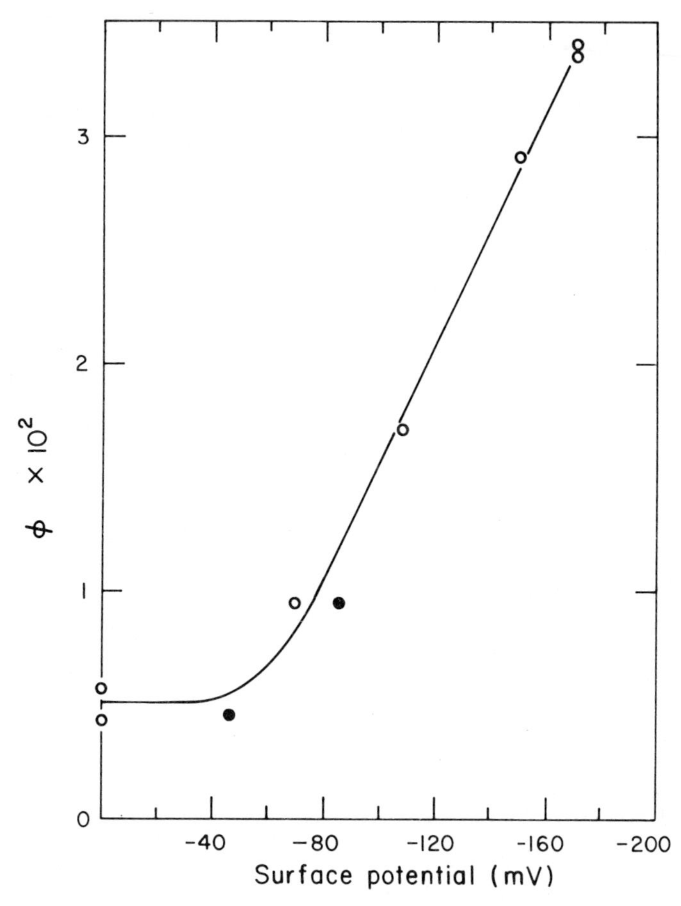

Figure 11. Quantum yield for propylviologen, $PVS^{\cdot -}$, formation as a function of the surface potential of negatively charged interfaces. Key: ○, SiO_2 system; and ●, NaLS micellar system (22).

Chemical Utilization of the Photoproducts in the Photodecomposition of Water

The different interfacial systems described in this paper represent supramolecular assemblies for the separation and stabilization of photoproducts. These photoproducts generated in the electron transfer reactions are an oxidized sensitizer and a reduced species (acceptor). In all systems that have been described here a sacrifical electron donor (EDTA or TEA) has been used. For any practical configuration, this sacrificial component must be excluded and water itself should be the compound oxidized. The oxidized intermediates, $Ru(bipy)_3^{3+}$ and Zn-TMPyP^{5+}, have the potential for oxidizing water to oxygen (E_o($Ru(bipy)_3^{3+}$/$Ru(bipy)_3^{2+}$) = 1.26 volt; E_o (Zn-TMPyP^{5+}/Zn-TMPyP^{4+}) = 1.2 volt) ,but since the reaction requires a concerted four-electron process while the photoproducts are single electron oxidants, a mediating charge-storage catalyst is needed. In recent years transition metal oxides and, in particular, RuO_2 and PtO_2, have been reported to act as oxygen evolution catalysts with $Ru(bipy)_3^{3+}$ as oxidant (23,24). In an analogous way, the reduced species produced in the photosensitized reaction should be coupled to hydrogen evolution. Reduced bipyridinium salts (viologen radicals) are capable of reducing water to hydrogen (25, 26). For this reaction colloidal platinum has been found to be an efficient charge-storage catalyst.

A schematic view of one possible complete system is shown in Figure 12. Since the oxidized photoproduct, e.g., $Ru(bipy)_3^{3+}$, is associated with one colloidal particle, its interface should be coated with an oxygen-evolving catalyst. An additional colloidal site is introduced by supporting platinum on a negatively charged polymer. The electrostatic repulsions of the two negatively charged interfaces would prevent agglomeration. By using a polymer with a low enough surface potential the approach of the reduced photoproduct, PVS$^{\cdot-}$, to the hydrogen evolution catalyst would be permitted while its recombination with $Ru(bipy)_3^{3+}$ on the other, more highly charged colloid, would be prevented. In this way, the vectorial character of the electron transfer process could be used for an efficient cleavage of water.

Acknowledgement: The work was supported, in part, by the Director, Office of Energy Research, Office of Basic Energy Sciences, Chemical Science Division, of the U.S. Department of Energy under Contract W-7405-ENG-48 and by the Netherlands Organization for the Advancement of Pure Research (Z.W.O.).

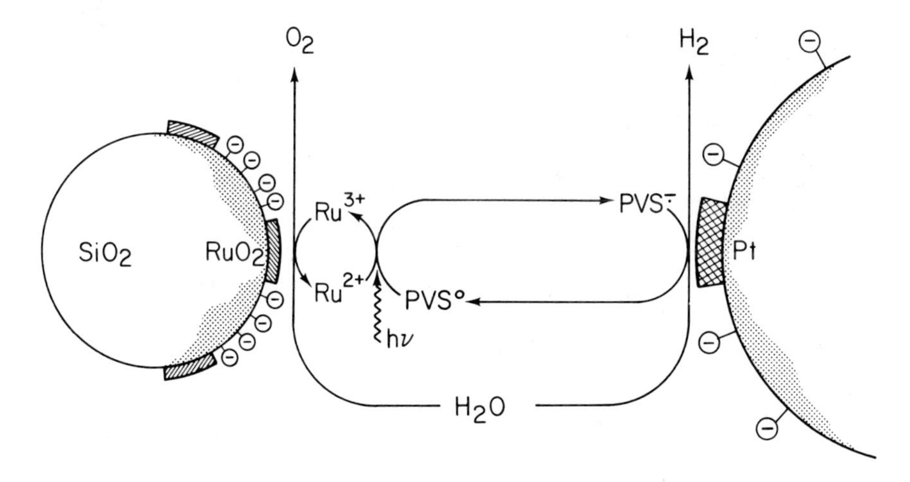

Figure 12. Utilization of SiO_2 colloids in the photodecomposition of water.

Literature Cited

1. Dolphin, D., McKenna, C, E., Murakami, Y., and Tabushi, I., Eds.; "Biomimetic Chemistry"; American Chemical Society, Washington, DC, 1981. Advances in Chemistry Series No. 191.
2. Breslow, R. Acc. Chem. Res. 1980, 13, 170; Isr. J. Chem. 1979, 18, 187.
3. Bolton, J. R. Science, 1978, 202, 105; Bard, A. J. Science, 1980, 207, 139; Porter, G. Pure and Appl. Chem. 1978, 50, 263.
4. Calvin, M. Acc. Chem. Res. 1978, 10, 369; Willner, I.; Ford, W. E.; Otvos, J. W.; Calvin, M. in "Bioelectrochemistry", Keyzer, H. and Gutmann, F. Eds.; Plenum Press, New York, 1980, pp. 558-581.
5. Gibbs, M. "Structure and Functions of Chloroplasts"; Springer-Verlag, Berlin, 1971; Calvin, M. and Bassham, J. A. "The Path of Carbon in Photosynthesis"; Prentice-Hall, Englewood Cliffs, 1957.
6. Balzani, V.; Boletti, M. T.; Gandolfi, M. T.; Maestri, M Top. Curr. Chem. 1978, 75, 1; Balzani, V.; Bolleta, F.; Scandola, F.; Ballardini, R. Pure and Appl. Chem. 1979, 51, 299.
7. Whitten, D. G. Rev. Chem. Intermediates, 1979, 2, 107; Hopf, R.; Whitten, D. G. in "Porphyrins and Metalloporphyrins", Smith, K. M. Ed.; Elsevier, New York, 1975, pp. 667-700.
8. Whitten, D. G. Acc. Chem. Res. 1980, 13, 83.
9. Sutin, N. J. Photochem. 1979, 10, 19.
10. Kalyanasundaram, K. Chem. Soc. Rev. 1978, 7, 453; Turro, N. J.; Grätzel, M.; Brown, A. M. Angew. Chem. Internat. Ed. 1980, 19, 573.
11. Brugger, P. A.; Grätzel, M. J. Amer. Chem. Soc. 1980, 102, 2461.
12. Jones, C. A.; Weaner, L. E.; Mackay, R. A. J. Phys. Chem. 1980, 84, 1495.
13. Willner, I.; Ford, W. E.; Otvos, J. W.; Calvin, M. Nature (London), 1979, 280, 823.
14. Ford, W. D.; Otvos, J. W.; Calvin, M. Nature (London), 1978 274, 507; Proc. Natl. Acad. Sci. U.S.A. 1979, 76, 3590.
15. Infelta, P.P.; Grätzel, M.; Fendler, J. H. J. Amer. Chem. Soc. 1980, 102, 1479; Matsuo, T.; Itoh, K.; Takuma, K.; Hashimoto, K.; Nagamura, T. Chem. Lett. 1980, 8, 1009.
16. Laane, C.; Ford, W. E.; Otvos,J. W.; Calvin, M. Proc. Natl. Acad. Sci. U.S.A., 1981, 78, 0000.
17. Iler, R. K. "The Colloid Chemistry of Silica and Silicates"; Cornell University Press, Ithaca, New York (1955); Loeb, A. L.; Overbeek, J.Th.G. Wiersema, P. H. "The Electrical Double Layer Around a Spherical Colloid Particle", M.I.T. Press, Cambridge, Massachusetts (1961).

18. James, A. M. Chem. Soc. Rev. 1979, 8, 289.
19. Willner, I.; Otvos, J. W.; Calvin, M. J. Amer. Chem. Soc. 1981, 103, 0000.
20. Willner, I.; Yang, J.-M.; Otvos, J. W.; Calvin, M. J. Amer. Chem. Soc. submitted for publication.
21. Pellizzetti, E.; Pramaura, E. Inorg. Chem. 1979, 18, 882.
22. Laane, C.; Willner, I.; Otvos, J. W.; Calvin, M. J. Amer. Chem. Soc. submitted for publication.
23. Lehn, J.-M.; Sauvage, J. P.; Ziessel, R. Nouv. J. Chim. 1979, 3, 423; 1980, 4, 623.
24. Borgarello, E.; Kiwi, J.; Pellizzetti, M. V.; Grätzel, M. Nature (London), 1981, 289, 158; Kalyanasundaram, K.; Micic, O.; Promauro, E.; Grätzel, M. Helv. Chim. Acta, 1979, 62, 2432.
25. Moradpour, Z.; Amoruyal, E.; Keller, P.; Kagan, H. Nouv. J. Chim. 1978, 2, 547; J. Amer. Chem. Soc. 1980, 102, 7193.
26. Kalyanasundaram, K.; Kiwi, J.; Grätzel, M. Helv. Chim. Acta, 1978, 61, 2720; DeLaive, D. J.; Sullivan, B. P.; Meyer, T.T.; Whitten, D. G. J. Amer. Chem. Soc. 1979, 101, 4007.

RECEIVED May 27, 1981.

6

Photochemistry on Colloidal Silica Solutions

J. WHEELER and J. K. THOMAS

University of Notre Dame, Department of Chemistry, Notre Dame, IN 46556

Two probe molecules, Ruthenium tris-bipyridyl, Ru(II), and 4-(1-pyrenyl)butyltrimethylammonium bromide, PN^+ have been used to investigate the nature of colloidal silica particles in water. The fluorescence spectra of the two probes show that the silica surface is very polar and similar to water. Quenching studies of the excited state of RuII and PN^+ by anionic quenching molecules show that the particles are negatively charged but that the charge is not as effective as that on sodium lauryl sulfate micelles. Quenching studies with cationic quenchers show that the cations are bound strongly to the silica particles but do not move as readily around the surface as on anionic micelles. A small steric effect is observed with neutral quenchers. Several charge transfer reactions, including photo-ionization are strongly affected by the silica particles. The studies show many similarities to anionic micelles; they differ from micelles in two important aspects: (a) they do not solubilize neutral organic molecules and (b) cationic organic molecules such as PN^+, hexadecyltrimethylammonium bromide, and hexadecylpyridinium chloride, tend to cluster on the silica surface rather than disperse uniformly around it as with ionic micelles.

The past decade has seen great strides in the utilization of organized assemblies, such as micelles, microemulsions, etc., to promote desirable features of photochemical reactions.[2,3] The most prominent feature of these systems is the use of an ionic surfactant, such as sodium lauryl sulfate, NaLS, or hexadecyltrimethylammonium bromide, CTAB, to create a charged barrier between the lipid and aqueous phases of a small region of the system. Reactants located

0097-6156/82/0177-0097$05.00/0

at this interface are strongly influenced in their subsequent reactions by: -

(a) close proximity for rapid reaction
(b) the strong electric field of the surface which influences electron transfer reactions, repeals ions of similar charge to the surface, and attracts ions of opposite charge
(c) Organization of reactants to produce specific desired effects
(d) solubilize hydrophobic molecules in close proximity to a hydrophillic surface.

Future developments in utilization of organized assemblies could lie in the use of colloidal semiconductors,(4) and the use of inorganic colloids in place of the organic surfactants indicated above. One particular system of interest is to use colloid Bentonite clays, which as they strongly promote thermal reactions, should also promote photochemical reactions.(5) However, a clay system is quite different than the simple micelles which have been studied, being a static very polar structure which can absorb cations by exchanging the present clay cations. The exact nature of adsorption of uncharged organic molecules is uncertain, but cations are expected to be strongly absorbed.

To initiate our studies we report data on photochemistry in aqueous solutions of silica. These solutions are mainly water and the silica particles possess a negative charge which is counteracted by an invisible sodium ion. The systems bear some resemblance of anionic micelles such as NaLS.

Experimental

Absorption spectra were recorded in a Perkin-Elmer spectrophotometer, and fluorescence spectra were recorded on a Perkin-Elmer 44B spectrofluorimeter. Flash photolysis studies were carried out using an excimer laser, λ excitation = 3080A°, a ruby laser, λ excitation = 3471A°, or a nitrogen laser, λ excitation = 3391A°. The system has been described previously.(6)

The colloidal silica solutions were obtained from NALCOAG Chemicals #1115, pH 10.4, r = 40A° radius; #1034-A, pH 3.2, r = 200A°; #1050, pH 9.0, r = 200A°. These solutions were run at 20% dilution in water.

A new probe molecule 4-(1-pyrenyl)butyltrimethylammonium bromide was made from pyrene butyric acid as follows: - pyrene butyric acid was refluxed with methanol and converted to the methyl ester,

$$\text{Py} - (CH_2)_3\text{-COOH} \xrightarrow{CH_3OH} \text{Py} - (CH_2)_3 - C{\overset{\displaystyle O}{\underset{\displaystyle OCH_3}{\lessgtr}}}$$

The ester was reduced with lithium aluminum hydride to the alcohol,

$$Py(CH_2)_3\text{-}\ CH_2OH \xrightarrow[\text{ether}]{LiAlH_4} Py(CH_2)_3\text{-}\ CH_2OH \qquad \text{mp } 81\text{-}83°C$$

The alcohol was converted to the bromide by refluxing with CBr_4 and triphenylphosphine:

$$Py(CH_2)_3\text{-}\ CH_2OH \xrightarrow[\phi_3P]{CBr_4} Py(CH_2)_3\text{-}CH_2Br \qquad \text{mp } 76\text{-}77°C$$

The bromide derivative was refluxed with trimethylamine to produce the quaternary ammonium salt, PN^+.

$$Py(CH_2)_3\text{-}CH_2\text{-}Br \xrightarrow{NMe_3} [Py(CH_2)_4\text{-}N(CH_3)_3]^+\ Br^-$$

Experimental Data and Discussion

Spectroscopy. It has been shown previously[7] that the fluorescence spectrum of Ruthenium tris-bipyridine, RuII, is solvent dependent, showing a red shift with increasing solvent polarity. The fluorescence spectrum of RuII on silica particles is identical to that of the excited molecule in water. It will be shown subsequently that the RuII is essentially all bound to the silica particle, hence the data show that the environment of a probe molecule such as RuII on silica particles is very polar and similar to water.

Essentially, the same data is obtained for the organic probe molecule 4-(1-pyrenyl)butyltrimethylammonium bromide, PN^+. The fluorescence spectra of this molecule in several environments including silica and NaLS micelles are shown in Fig. 1. The water and silica spectra are identical, thus confirming the RuII probe data.

Kinetic Data: RuII System

Fig. 2 shows the rate of formation and subsequent decay of excited RuII following laser excitation. The first order plot of the data is also shown, and the slope of this linear plot gives the rate constant for the process and the half line of reaction. Addition of a quenching molecule to the solution increases the rate of decay of (RuII)*: -

$$\text{RuII} \xrightarrow{h\nu} (\text{RuII})^*$$

$$(\text{RuII})^* \xrightarrow{k_1} \text{RuII} + h\nu \qquad (\text{RuII})^* \xrightarrow[k_2]{+Q} \text{products}$$

The concentration of quencher [Q], is much larger than [(RuII)*] so reaction 2 is pseudo first order. The overall rate constant k_1 for decay of (RuII)* from data, such as those

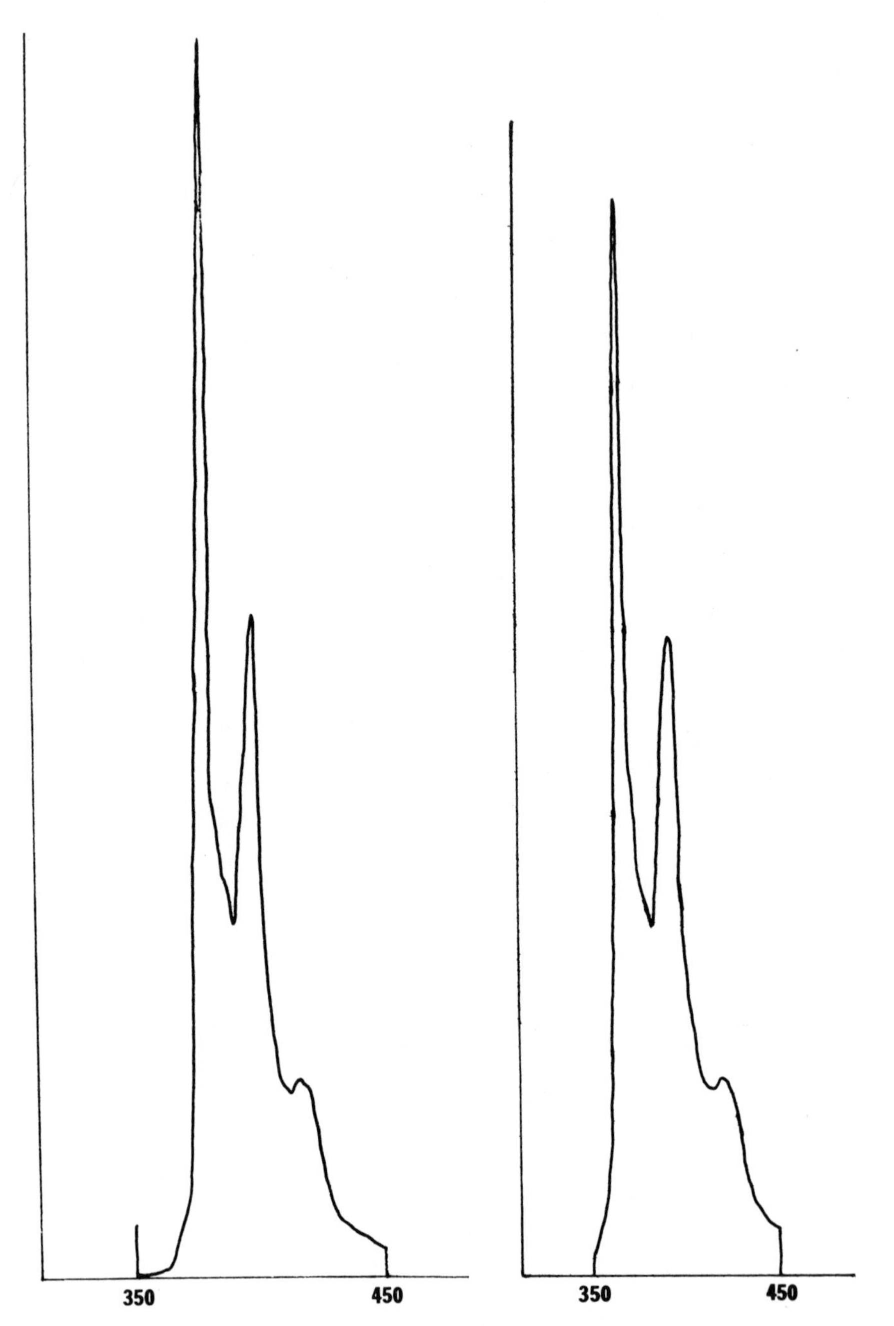

Figure 1a. Fluorescence spectra of PN^+ (2×10^{-6} M) from 350 to 450 nm in water. $\lambda_{ex} = 340$ nm.

Figure 1b. Fluorescence spectra of PN^+ (2×10^{-6} M) from 350 to 450 nm in 20% silica (#1115). $\lambda_{ex} = 340$ nm.

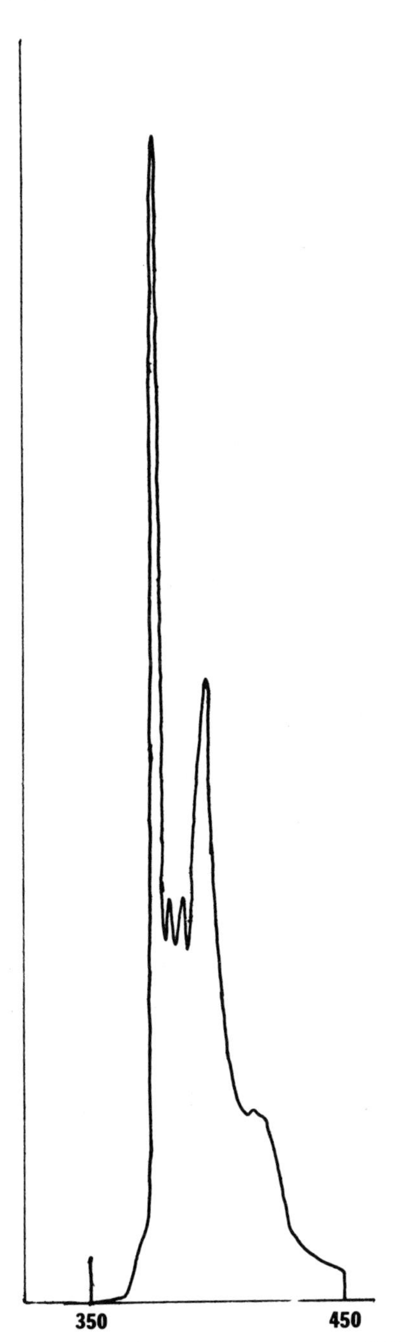

Figure 1c. Fluorescence spectra of PN^+ (2×10^{-6} M) from 350 to 450 nm in 0.5 M NaLS. $\lambda_{ex} = 340$ nm.

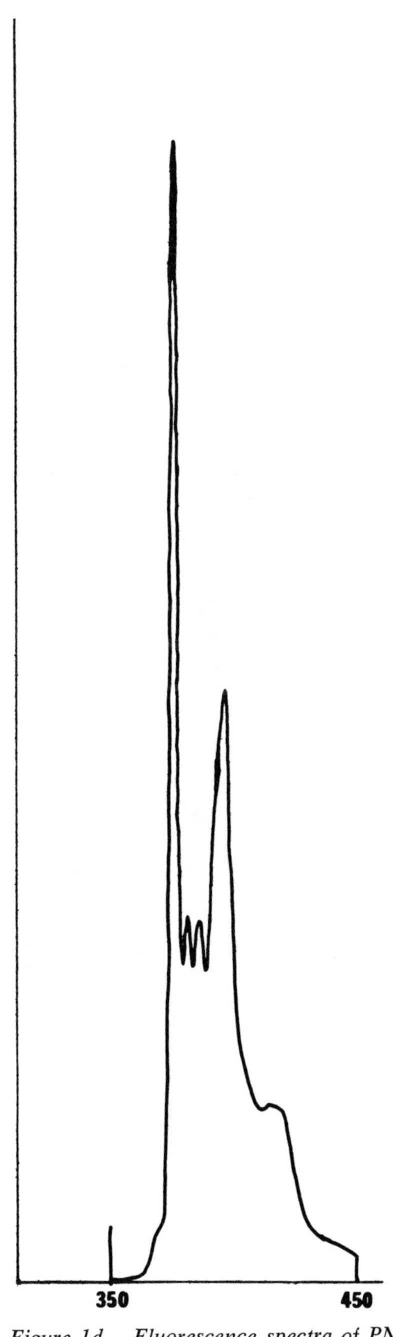

Figure 1d. Fluorescence spectra of PN^+ (2×10^{-6} M) from 350 to 450 nm in methanol. $\lambda_{ex} = 340$ nm.

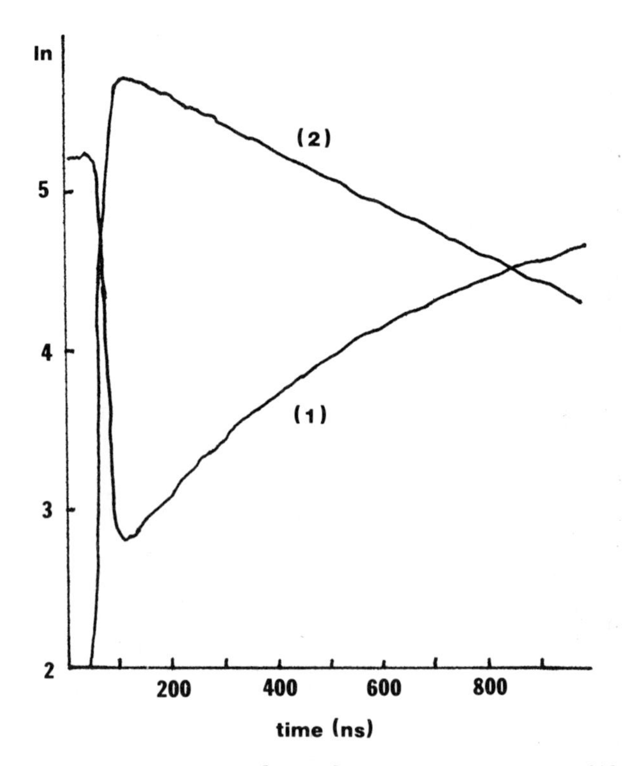

Figure 2. Fluorescence decay of [Ru(II)] in water where λ_{em} = 610 nm. Key: 1, fluorescence vs. time in arbitrary units; and 2, natural log of fluorescence vs. time; rate = 1.71×10^6 s^{-1}; half-life = 414 ns.*

shown in Figure 2 is thus: $k = k_1 + k_2 [Q]$. Several rate constants for reaction of (RuII)* with quenchers are shown in Table I. The data may be divided into three sections, quenching by uncharged species, quenching by anions, and quenching by cations. The rate constants for reaction in silica systems are compared to data for these systems in homogeneous solution i.e. water, and in anionic NaLS micelles. It is noted that the quenching rates for O_2 and nitrobenzene is about 25% smaller than those in water, but similar to those in NaLS. This is attributed to a steric factor imposed by the silica background on the approach of Q to (RuII)*.

Both negative ions $Fe(CN)_6^{3-}$ and 3-5 dinitrobenzoate are much slower in silica compared to water but not as slow as in NaLS where the rates were too slow to measure accurately. The decreased reaction rates are due to repulsion of the anionic quencher by the anionic silica particles on NaLS micelles.[8] It is possible to use the Debye modification of the Smoluckowski equation to explain these data.[8,9] The equation indicates that the diffusion controlled rate constant k_1 for reaction of two ions is given by

$$k = \frac{4\Pi r\ D\ N}{1000}\ \frac{Z_1 Z_2 e^2}{rEkT} \Bigg/ \left[\exp\left\{ \frac{Z_1 Z_2 e^2}{rEkT} \right\} - 1 \right]$$

where N is an Avagadro's number, r is the interaction radius, E the dielectric constant of the medium and Z_1e and Z_2e are the charges of the two reactants and D is the total diffusion constant. The RuII data can be explained quantitatively if charges of -8 to -10 units/particle are used for silica; a much larger charge (>20), is necessary to explain the NaLS data. The exact position of the probe in the surface is important in this calculation, and the data could indicate that for reaction to occur the anionic quenchers do not have to penetrate as close to SiO_2 as to NaLS.

The cation Cu^{++} is strongly bound to the silica particles and to NaLS micelles.[10] Particles or micelles containing (RuII)* and Cu^{++} should show quenching of (RuII)* that is more rapid than that in water. This is the case for NaLS micelles but the rates on silica are actually lower than those observed in water. This is due to a lower mobility of Cu^{++} around a silica particle compared to an NaLS micelle, a fact already indicated by the strong binding of ions such as RuII to silica particles.

Kinetic Studies, PN^+ System

Table II shows the quenching rate constants for excited PN^+ with several quencher molecules on silica and on NaLS micelles. The patterns shown by neutral quenchers e.g. O_2, CH_3NO_2 and dimethylaniline, anionic quenchers e.g. 3-nitropro-

Table I Ru II, observed at λ = 610 nm

	System, k ($LM^{-1}S^{-1}$)		
Quencher	Water	NaLS, micelle	Silica
None	1.67×10^6	1.3×10^6	1.4×10^6 (1.5×10^6) $[1.5 \times 10^6]$
O_2	2.8×10^9	2.1×10^9	1.3×10^9 $[1.2 \times 10^9]$
Nitrobenzene	3×10^9	1×10^9	7.6×10^8 (8.8×10^8) $[7.5 \times 10^8]$
$Fe(CN)_6^{3-}$	3.8×10^{10}	$<10^7$	1.0×10^8 (3×10^9) $[1.0 \times 10^8]$
3,5-dinitrobenzoate	7.8×10^9	$<10^7$	1.5×10^8
Cu^{++}	5.3×10^7	1.8×10^8	(1.6×10^7)
Heptyl Viologen	7.4×10^8	8.75×10^8	3.43×10^8 (3.6×10^8)

- Small pH 10.4 r = 40A°   ( ) acidic r = 200A°   [ ] Large pH 9.0 r = 200A°

Table II PN^+, observed at λ = 375 nm

Quencher	System k ($LM^{-1}S^{-1}$)		
	Water	NaLS, micelle	Silica
None	6.8×10^6	4.9×10^6	7.06×10^6
O_2	1.0×10^{10}	7.0×10^9	1.8×10^9
CH_3NO_2	3.8×10^9	2.6×10^9	1.1×10^9
Dimethylaniline	4.0×10^9	(Poisson)	1.1×10^9
Tl^+	3.2×10^9	1.7×10^{10}	1.6×10^{10}
Cu^{++}	1.947×10^9	1.8×10^9	5.4×10^8 (2% acidic Si)
3-nitropropionate	6.7×10^9	$<10^7$	2.0×10^8

pionate, and positive quenchers e.g. Tl^+ and Cu^{++} are quite similar to the (RuII)*, and are interpreted in a similar fashion.

At higher concentrations (PN^+) forms excimers, the rate of formation of which is too fast to be measured by the equipment (resolution <5 nsec). $(PN^+)^* + PN^+ \longrightarrow (PN^+)_2^*$. PN^+ is also quenched by hexadecylpyridinium ions on the silica particles. This quenching rate is also too fast to be measured. One hexadecylpyridinium ion CP^+ is sufficient to totally quench $(PN^+)^*$. The quenching is static in nature i.e. the yield of excited PN^+ decreases with increasing PN^+ ($(PN^+)_2$ being formed), or CP^+, while the fluorescent lifetime of PN^+ is unaffected. Such effects are not observed in micellar systems, but diffusion controlled movement of the reactants is required for reaction. It is suggested that the organic cation only possess an affinity for the silica particles via electrostatic attraction. This same attractive force is also operative in micelles, but in addition the hydrophobic environment of the carbon micelle provides an additional solubilization force which disperses the organic cations randomly around the micelle. This dispersion force is not operative in silica particles and the organic cations face an aqueous environment only as they cannot penetrate into the solid silica particle. If two organic cations are placed in a silica particle, then a stable and favorable configuration is attained if the molecules group together in close proximity and eliminate water from the hydrocarbon structure. This "colony" type process leads to static reaction, producing $(PN^+)_2^*$ or quenching of $(PN^+)^*$ by CP^+.

<u>Intermediates Produced (PN^+) System</u>

Fig. 3 shows the spectra of short lived species produced in the laser flash photolysis of PN^+ in water, NaLS, and in silica particles. Previous work enable us to identify the excited triplet and singlet states of the pyrene moiety, together with the pyrene cation and hydrated electron, e^-_{aq}

excited pyrene singlet PN^S 4700A°
excited pyrene triplet PN^T 4150A°
pyrene cation PN^{++} 4600A°
e^-_{aq} 7200A°; using absorption to the red

The PN^S is readily identified as its decay is that of the PN^+ fluorescence; PN^T is long lived and reactive with O_2 but not with N_2O, PN^{++} is very long lived and unreactive with O_2 or N_2O, while e^-_{aq} is short lived and very reactive with O_2 and N_2O. The photo-ionization process giving rise to PN^{++} and e^-_{aq}

$$PN^+ \longrightarrow PN^{++} + e^-_{aq}$$

is two photon and depends on the square of the laser intensity [14].

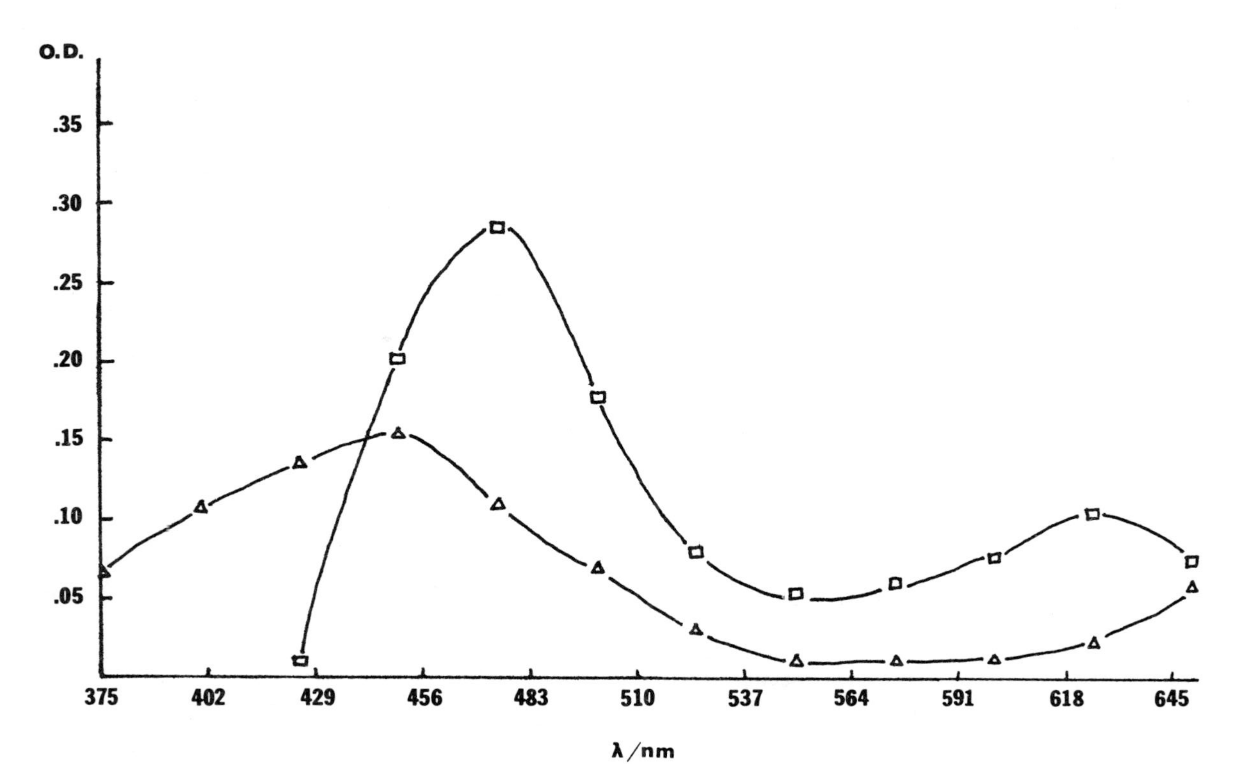

Figure 3a. Short-lived intermediates in the laser flash absorption photolysis of 10^{-4} M PN^+ in water. Key: □, *end of pulse; and* △, *0.6 μs after the pulse for water and 3 μs after the pulse for silica and NaLS.*

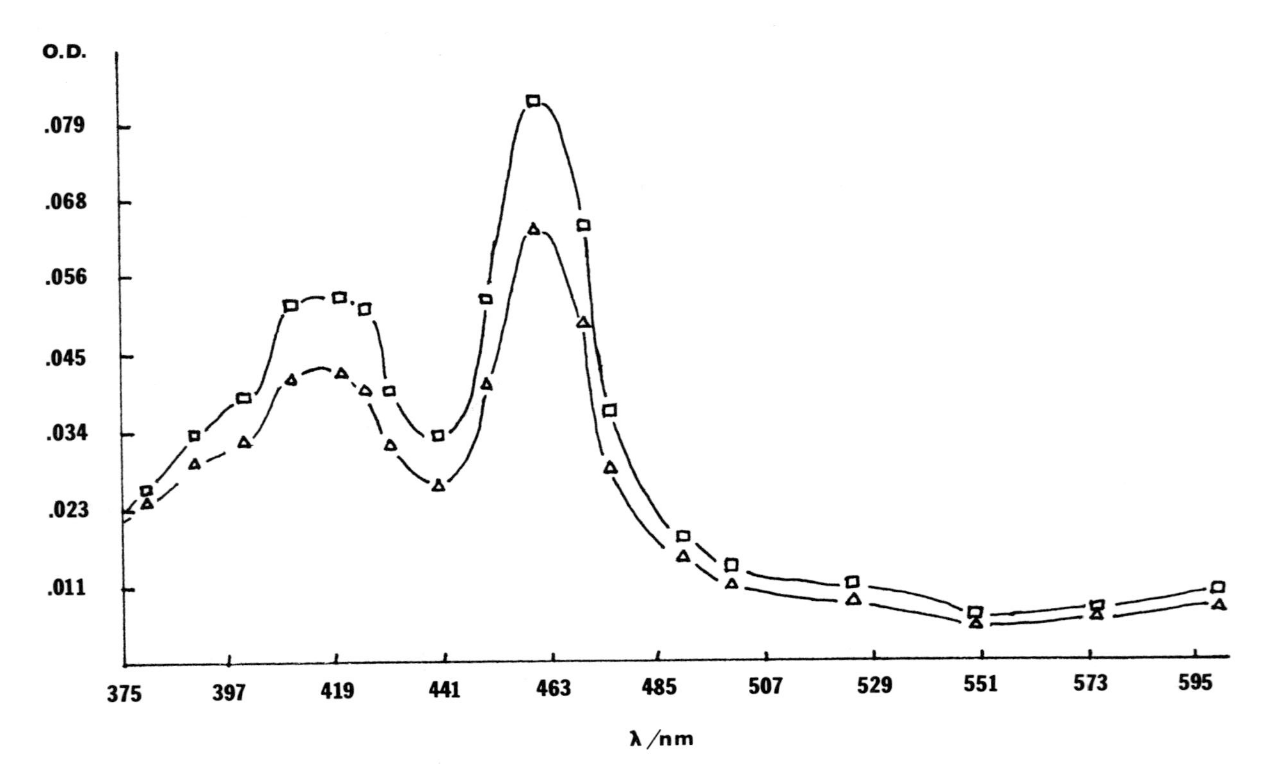

Figure 3b. Short-lived intermediates in the laser flash absorption photolysis of 10^{-4} M PN^{+} *in 20% silica (#1115). Key as in Figure 3a.*

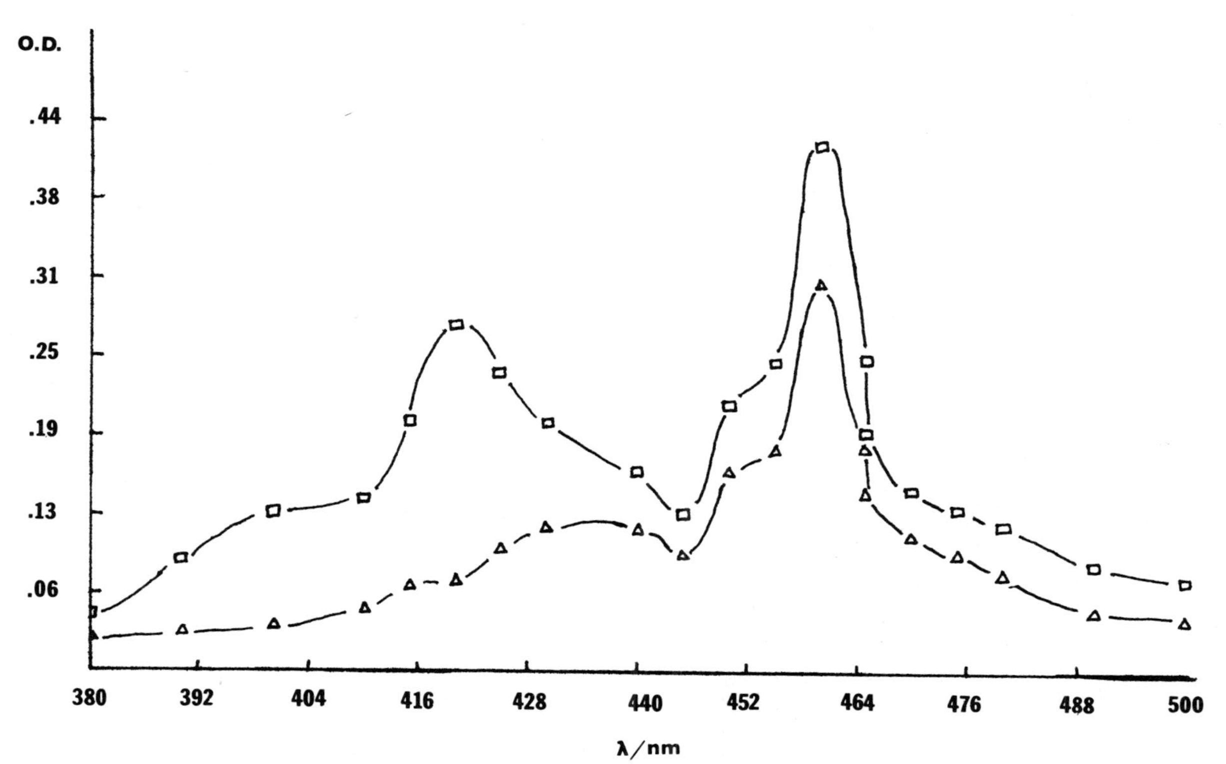

Figure 3c. Short-lived intermediates in the laser flash absorption photolysis of 10^{-4} M PN^{+} in 0.1 M NaLS. Key as in Figure 3a.

The lifetime of PN^{++} is much enhanced on the anionic NaLS and silica particles compared to water. An interesting feature of the silica particles is that PN^{++} is easily observed, but unlike micelles or water e^-_{aq} is not observed. This is not due to the fact that e^-_{aq} reacts with silica, as e^-_{aq} produced in the water bulk by photo-ionization of pyrene sulfonic acid (this molecule does not well bind to silica), has a long lifetime, >2 μsec. It is suggested that the photo-ionized e^- is ejected into the silica particle where it is stabilized and not observed over the spectral range studied, λ = 3000A° to 6500A°.

Dimethylaniline rapidly quenches $(PN^+)^*$ on silica and produces the anion of PN^+, $(PN^+)^-$ and the DMA cation, DMA^+. The ions are short lived as DMA^+ and $(PN^+)^-$ do not escape from the particle rapidly enough to prevent back electron transfer. This has also been observed in anionic NaLS micelles.(14)

Laser excitation of RuII leads to (RuII)* and to a bleaching of the RuII ground state absorption in the region λ~4600A°. Heptyl and methyl viologen, HV^{++}, MV^{++}, rapidly quench the (RuII)* but the well established release of long lived intermediates such as reduced MV^+ is not observed; (RuII)* + MV^{++} ⟶ (RuIII) + MV^+. This is similar to the PN^+ - DMA system where the anionic silica surface binds the cationic products and promotes back e^- transfer before the product ions can be separated.

It is interesting to note that Ag^+ reacts with (RuII)* on silica leading to a long lived, (several seconds) bleaching of RuII and to the formation of colloidal silver. The reaction is:

$$(\text{RuII})^* + Ag^+ \longrightarrow (\text{RuIII}) + Ag^\circ \quad \text{Colloidal silver}$$

The back reaction of Ag° + (RuIII) is rapid in water, but is strongly retarded in silica where Ag° is ejected from the vicinity of (RuIII) which is strongly bound to the silica particle. Such a long lived separation of products is not observed in homogeneous aqueous solution.

Conclusion

The data show that in many ways silica particles behave in a similar fashion to anionic micelles, although the quantitative aspects of the anionic surface are different. A major difference occurs in the location of cationic organic molecules on the particles, as these molecules tend to cluster together rather than disperse uniformly as in micellar systems.

Literature Cited

1. The authors wish to thank the Army Research Office via grant No. DAAG29-80-K-0007, P001, for support of this research.
2. Turro, N. J.; Grätzel, M.; Braun, A. M. Angewardte Chemi. 1980, 19, 675.
3. Thomas, J. K. Chem. Rev. 1980, 80, 283.
4. Harbour, J. R.; Hair, M. L. J. Phys. Chem. 1978, 82, 1397.
5. Grim, R. E.; Clay Mineralogy 1968, McGraw Hill, N. Y.
6. McNeil, R.; Richards, J. T.; Thomas, J. K. J. Phys. Chem. 1970, 74, 2290. Also Atik, S. S.; Thomas, J. K. JACS in Press.
7. Meisels, D.; Matheson, M.; Rabani, J. JACS 1978, 100, 117.
8. Thomas, J. K.; Accounts of Chem. Research 1977, 10, 133.
9. Matheson, M. S. Solvated Electron ACS Advances in Chem. 1965, No. 50, p 45.
10. Grätzel, M.; Thomas, J. K. J. Phys. Chem 1974, 78, 2248.
11. Atik, S. S.; Thomas, J. K. JACS in Press.
12. Richards, J. T.; West, G.; Thomas, J. K. J. Phys. Chem. 1970, 74, 4137.
13. Wallace, S. C.; Thomas, J. K. Rad. Res. 1972, 10, 76.
14. Katusin-Razem, B.; Wong, M.; Thomas, J. K. JACS 1978, 100, 19679.

RECEIVED August 4, 1981.

7

Efficient Water Cleavage by Visible Light in Colloidal Solutions of Bifunctional Redox Catalysts

MICHAEL GRÄTZEL

Ecole Polytechnique Fédérale de Lausanne, Institut de Chimie Physique, 1015 Lausanne, Switzerland

Cleavage of water by four quanta of visible light into hydrogen and oxygen is achieved in aqueous solutions of RuO_2 and Pt cosupported by colloidal TiO_2 particles. The only other component present is a sensitizer. No electron relay compound is required to accomplish the photolysis. Amphiphilic surfactant derivatives of $Ru(bipy)_3^{2+}$ exhibit astonishingly high activity in promoting the water cleavage process. Adsorption of the sensitizer at the TiO_2 surface is evoked to explain the observations. Exposure to UV radiation leads to efficient water cleavage in the absence of a sensitizer.

In the present area of dwindling fuel reserves the development of alternative energy supplies has become a research subject of high priority (1-7). One topic that has intrigued scientists from many different fields is the photolysis of water using solar radiation (8-15). The practical potential of devices achieving this process would be enormous if sufficiently high conversion efficiencies could be obtained. Figure 1 shows a plot of the maximum conversion efficiency of a threshold absorber that would use all the photons in the solar spectrum below an onset wavelength λ. In view of the thermodynamic requirements for the splitting of water into hydrogen and oxygen by a 4 photon process

$$2H_2O \xrightarrow{4h\nu} 2H_2 + O_2 \quad (\Delta 6 = 1.23eV \simeq 1000) \qquad (1)$$

and unavoidable losses, a threshold wavelength of 600nm is probably a realistic estimate for a water cleavage device. This would correspond to an upper limit in the conversion efficiency of ca. 20%. According to more conservative estimates a 16% yield appears to be feasible (16). Still, such systems could make a tremendous contribution to satisfy future energy demands.

0097-6156/82/0177-0113$06.25/0

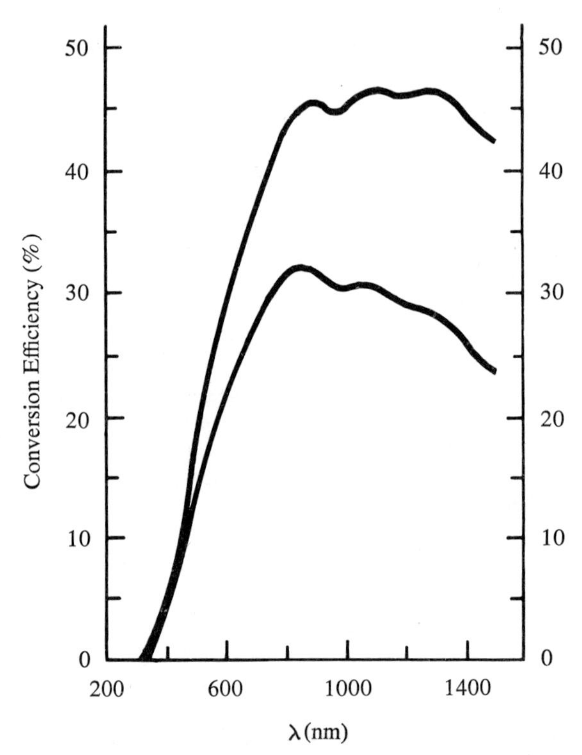

Figure 1. Optical conversion efficiency for solar energy as a function of threshold wavelength of the absorber. Curve I is a plot of the fraction of incident solar power (percent) available at various threshold wavelengths; Curve II is a plot of the thermodynamic conversion efficiencies under optimal rates of energy conversion.

In order to achieve the goal of decomposing water into hydrogen and oxygen by visible light, different strategies have been applied. A straightforward solution which suffers from high cost would be the coupling of a photovoltaic device with a water electrolyzer. An alternative approach is based on the concept of wet photovoltaics, i.e., the illumination of a semiconductor/electrolyte junction. Honda and collaborators have shown that photoelectrolysis of water can indeed be achieved with TiO_2 as a photoanode (17). However, this material absorbs only UV light, which makes it unsuitable for solar application. This lecture deals with yet another concept of photoinduced water cleavage based on microheterogeneous photochemistry and redox catalysis. Assemblies of minute dimensions will be described which through suitable molecular engineering accomplish the difficult task of light-induced charge separation coupled to energy conversion.

Among the variety of systems presently under investigation one can distinguish three different categories. The first comprises photosynthetic hybrid systems, Fig. 2. Here, chloroplasts or individual photosystems are employed as light harvesting units. The objective is to exploit the high efficiency of the primary photosynthetic redox events without attempting to synthesize carbohydrates from CO_2. Instead hydrogen generation from water is achieved by intercepting the electron transfer chain of photosystem I with a suitable electron relay such as methyl viologen (MV^{2+}). The latter couples the photosynthetic electron flow to a catalyst affording water reduction to hydrogen. Great progress has recently been made in this domain with the advent of synthetic catalysts such as ultrafine Pt particles (18) or Pt on PtO_2 (19) which replace the natural enzyme hydrogenase. Water oxidation is carried out by the water splitting enzyme of photosystem II. Attemps are also made to replace the latter by highly active oxygen generating catalysts such as colloidal RuO_2 (20).

The second approach is to employ synthetic molecular assemblies such as micelles, microemulsions and vesicles as reaction systems. Structural features of these molecular organizations are outlined in Figure 3. These aggregates simulate the microenvironment present in biological systems. Hydrophobic host molecules participating in photoredox reactions may be incorporated in their lipid-like interior and the charged lipid/water interface may be exploited to control kinetically the electron transfer events. Thus, in the case of an ionic micelle, the local electrostatic field present at the surface of the aggregate can readily exceed 10^6 V/cm and this microscopic electrostatic barrier can be used to achieve light-induced charge separation (21). The situation is analogous to electron-hole separation in the space charge layer of a semiconductor. In the case of the photoinitiated

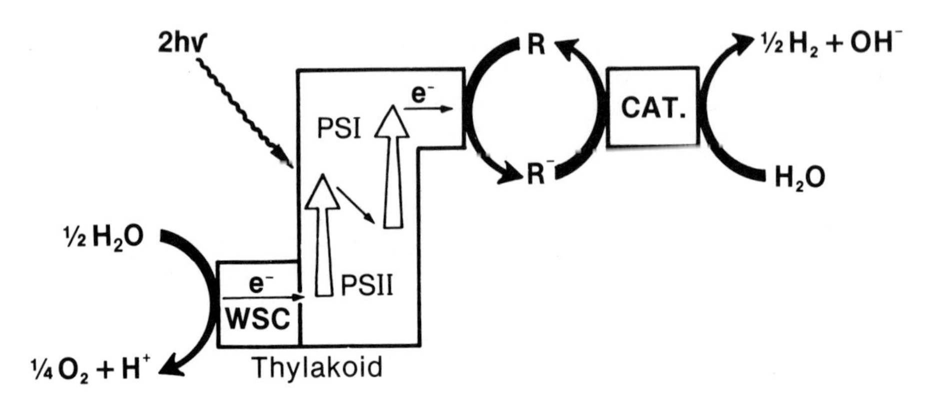

Figure 2. Schematic illustration of a cell-free hybrid system for the biophotolysis of water.

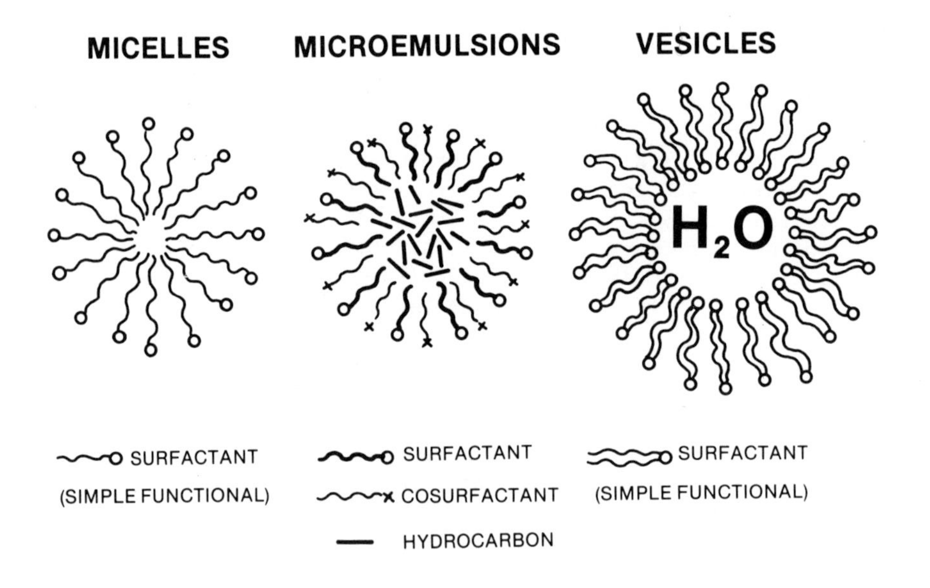

Figure 3. Structural features of colloidal assemblies employed in light-induced charge separation.

redox reaction between a sensitizer (S) and an electron relay (R) the goal is to enhance the rate of the forward reaction

$$S + R \underset{}{\overset{h\nu}{\rightleftharpoons}} S^{+} + R^{-} \tag{2}$$

and at the same time retard that of the backward electron transfer. This can be achieved by using solutions of simple (22) or functional (23) micellar aggregates as reaction medium. The latter type of surfactants are distinguished by the fact that they are chemically linked to moieties participating themselves in the redox events. Amphiphilic redox relays have also been successfully employed for light-induced charge separation. Here the addition of an electron to the molecule drastically shifts its hydrophobic/hydrophilic balance.

Apart from micelles, molecular assemblies such as microemulsions (24) and vesicles (25) deserve particular attention in the context of photoredox reactions in biomimetic aggregates. Calvin and co-workers have for the first time illustrated light-induced electron transfer across the bilayer of liposomes (26). The same group has also performed elegant studies with inverted micelles. The whole domain of biomimetic systems is experiencing presently a rapid growth and exciting discoveries manifest the astonishing progress in this area.

Light Harvesting Units in Artificial Water Splitting Systems

A third category which subsequently will be treated in more detail is that of totally artificial systems. These show no apparent similarity with their natural counterpart. Figure 4 summarizes the mechanism of light harvesting and energy conversion operative in three different configurations typically employed here. A sensitizer/relay pair is used in the first device. Light-induced electron transfer produces the radical ions S^{+} and R^{-} which are subsequently employed to oxidize and reduce water, respectively. Thus, light functions here as an electron pump operating against a gradient of chemical potential. Such photoinduced redox reactions have been studied in great detail both from the experimental (27) as well as theoretical point of view. The rate of electron transfer between excited sensitizer and relay is expected to approach the diffusion controlled limit as soon as the driving force for the reaction exceeds a few hundred millivolts. Conversely, the backward electron transfer between S^{+} and R^{-} which is thermodynamically strongly favored is almost always diffusion controlled. This poses a severe problem for the use of such systems in energy conversion devices as it limits the lifetime of the radical ions to at most several milliseconds under solar light intensity.

a/ SENSITIZER / RELAY PAIR

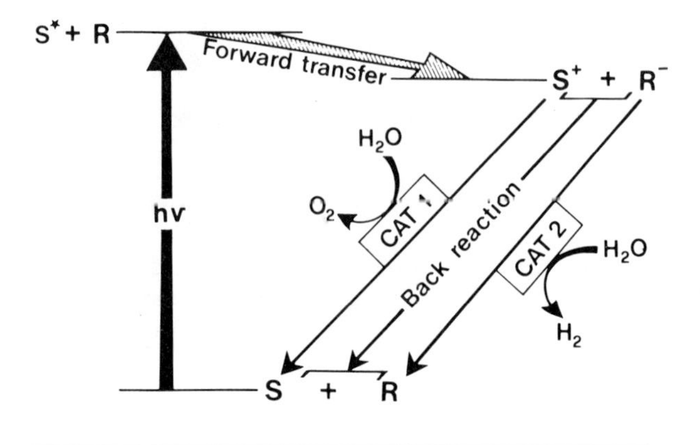

b/ SENSITIZER / COLLOIDAL SEMICONDUCTOR

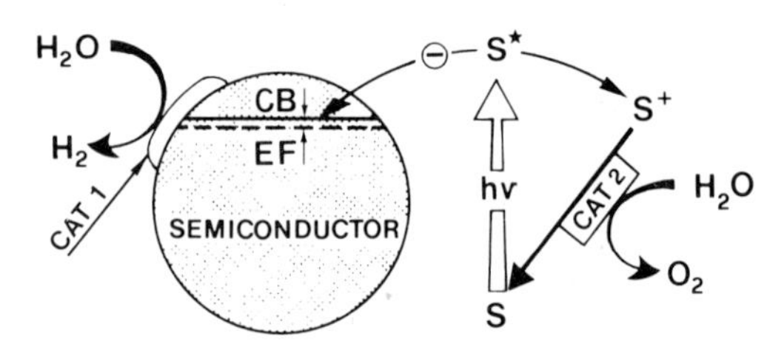

c/ COLLOIDAL SEMICONDUCTOR

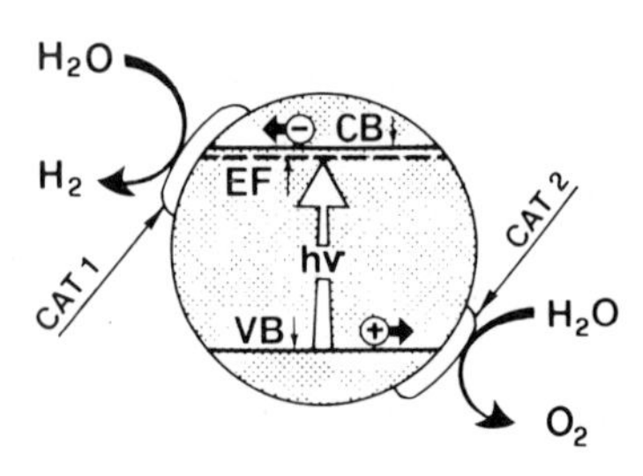

Figure 4. Light harvesting and catalytic units for light-induced water decomposition.

With regards to the choice of the sensitizer and the relay, compounds have to be found that are suitable both from the viewpoints of light absorption and redox potentials and undergo no chemical side reactions in the oxidation states of interest. The sensitizer should have good absorption features with respect to the solar spectrum. Also, its excited state should be formed with high quantum yield, have a reasonably long lifetime and the electron transfer reaction must occur with high efficiency, viz. good solvent cage escape yield of the redox products. The redox properties of the donor-acceptor relay must obviously be tuned to the fuel-producing transformation envisaged. In the case of water cleavage by light the thermodynamic requirements are such that $E^0(S^+/S) > 1.23V$ (NHE/ and $E^0(R/R^-) > 0V$ under standard conditions. In the design of sensitizer/relay couples suitable for photoinduced water decomposition considerable progress has been made over the last few years (28). A number of systems have been explored converting more than 90% of the threshold light energy required for excitation of the sensitizer into chemical potential. Porphyrines appear to be particularly promising in this respect. In several cases the reduced relay and oxidized sensitizer are thermodynamically capable of generating H_2 and O_2 from water:

$$R^- + H_2O \rightarrow \tfrac{1}{2} H_2 + OH^- + R \qquad (3)$$

$$2S^+ + H_2O \rightarrow \tfrac{1}{2} O_2 + 2H^+ + 2S \qquad (4)$$

Noteworthy examples are sensitizers such as $Ru(bipy)_3^{2+}$ (29), porphyrine derivatives (30) and acridine dyes, e.g., proflavin (31). Among the electron relay compounds investigated it is worth mentioning the viologens (32) Eu^{3+}, V^{3+} and their respective salicylate complexes (33), $Ru(bipy)_3^{3+}$ (34) and cobalt complexes (35). Thus, at present a considerable choice of sensitizer/relay pairs is available that fulfill the photochemical and thermodynamic requirements for water decomposition.

The next step is to combine the photoprocess with the generation of hydrogen using water as an electron source. This presents a formidable problem since water oxidation as well as reduction are multistep electron transfer processes that proceed through stages of highly reactive and energetic intermediates. The success in this domain has therefore largely been determined by the development of redox catalysts mediating hydrogen and oxygen formation and thus avoiding these radical intermediates. In fact, only through drastic improvement of previously known and discovery of new redox catalysts has the design of a cyclic water decomposition system operating on four quanta of visible light become feasible. The performance of these catalysts has to satisfy the following conditions: a) The catalysts have to intercept the thermal back reaction which occurs in the micro-to-millisecond time domain. b) The water reduction catalyst 2 must compete with oxygen reduction by R^- which is expected to occur at a diffusion controlled rate. This sets the rate limit required for H_2 generation at several microseconds. c) The intervention of the catalysts has to be specific in order to avoid short-circuitry of the back reaction. d) In order to achieve high quantum yields it is beneficial to keep the oxygen concentration in solution as low as possible. Hence, it is desirable to have an oxygen carrier present in solution that absorbs the O_2 produced during photolysis. This allows also for H_2/O_2 separation.

Development of Highly Efficient Hydrogen Producing Catalysts

A few years ago it would have seemed impossible to overcome all of the difficulties associated with requirements (a-d). At that time there was even no O_2 producing catalyst available. This was discovered in 1978 in our laboratory (36) in the form of noble metal oxides such as PtO_2, IrO_2 and RuO_2. The latter has been most widely investigated since then (37). Platinum has been known for a long time to mediate water reduction by agents such as V^{2+}, Cr^{2+} and also reduced viologens. However, it required several years of research to develop a Pt catalyst that would satisfy conditions (a-d) indicated above, i.e., produce H_2 in the microsecond time domain at reasonably low Pt concentrations.

Our strategy here was to develop Pt colloids of minimum particle size which would render the water reduction by the electron relay R^- particularly effective. The goal was to obtain hydrogen evolution rates in the microsecond time domain which would compete with back electron transfer. First a centrifuged Pt sol protected by polyvinyl alcohol was employed which gave clear and almost colorless aqueous suspensions even at high Pt concentrations (38). Flash photolysis technique was applied for the first time to study the dynamics of intervention of the Pt particles in the water reduction process. Methylviologen (MV^{2+}) was reduced by the excited state of $Ru(bipy)_3^{2+}$:

$$*Ru(bipy)_3^{2+} + MV^{2+} \rightarrow MV^{+} + Ru(bipy)_3^{3+} \quad (5)$$

and the behavior of MV^+ analyzed by absorbance technique. The rate constant for water reduction:

$$MV^{+} + H_2O \xrightarrow{Pt} \tfrac{1}{2} H_2 + OH^{-} + MV^{2+} \quad (6)$$

was found to be $6 \times 10^4 S^{-1}$ at 10^{-3}M Pt concentration. The high activity of this catalyst was also demonstrated in a classical photochemical reaction, i.e., the photolysis of benzophenone in a water/alcohol mixture. Here in the presence of colloidal Pt the ketyl radicals were directed towards water reduction (39)

$$\phi_2C-OH + H_2O \xrightarrow{Pt} \phi_2C=O + \tfrac{1}{2} H_2 + H_2O \qquad (7)$$

instead of dimerization or disproportionation. Thus the pathway of the photoreaction is totally altered in the presence of these catalysts. Instead of the photoreduction of benzophenone to benzpinakol one observes sensitized conversion of isopropanol into acetone and hydrogen, which is an energy storing process.

For further refinement of the Pt catalyst we made use of Turkevich's method to prepare ultrafine and monodisperse Pt sols (40). Employing polymer surfactants as protective agents a first successful attempt was made to render the intervention of these particles specific. Thus polyvinylpyridine absorbs well to the 30 Å platinum particles rendering their surface amphiphilic. Hydrophobic relays such as long chain substituted viologen radicals are readily trapped by these particles and subsequently affect water reduction. In contrast the oxidized sensitizer is rejected from the Pt surface by hydrophobic and electrostatic interactions. Thus, short-circuitry of the back reaction is avoided.

A further remarkable increase in activity is observed when these ultrafine Pt aggregates are deposited on a colloidal semiconductor such as TiO_2. In this case the cross-section of the reaction of the reduced relay R^- with the catalyst is greatly increased as the support itself can act as an electron acceptor (Figure 5). The electron is injected into the conduction band of the semiconductor from where it migrates to Pt sites where hydrogen generation from water takes place. By using laser photolysis technique, it was possible to directly monitor the kinetics of hydrogen formation from reduced methyl viologen and water in such a system (41). We observed a reaction rate constant of k = $3 \times 10^5 S^{-1}$ at 20mg Pt/l. The kinetic events are illustrated in Figure 6, which illustrates the time course of MV^+ decay (602nm) and $Ru(bipy)_3^{2+}$ bleaching (470nm) after laser excitation. In the absence of catalyst there is back reaction between these two species within a time domain of several hundred microseconds. Introduction of the Pt/TiO_2 catalyst in the solution causes a dramatic increase in the rate of MV^+ decay, which is due to charge injection into the colloidal TiO_2 and subsequent hydrogen formation. In contrast the $Ru(bipy)_3^{2+}$ bleaching is retarded in the presence of the catalyst showing that the particles interact selectively with the MV^+, while no or very slow reaction occurred

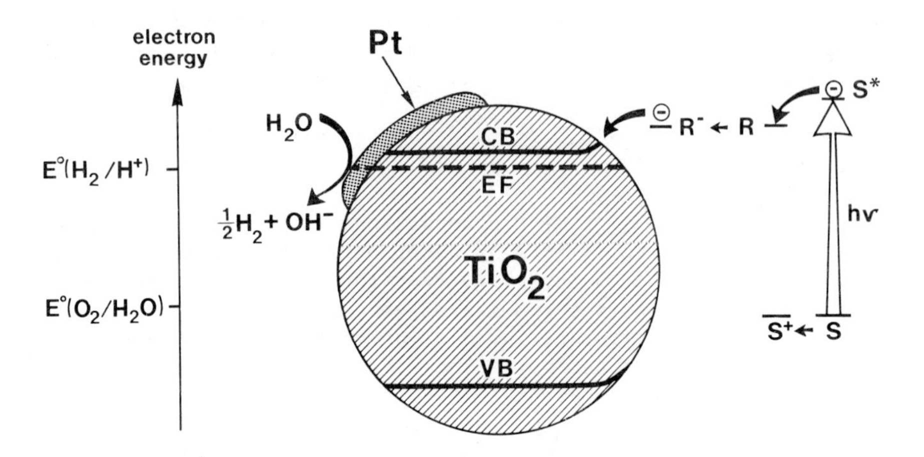

Figure 5. Mediator function of the colloidal TiO_2 particle, loaded with Pt in the light-induced H_2 generation from water. Electron injection from the reduced relay into the TiO_2 conduction band.

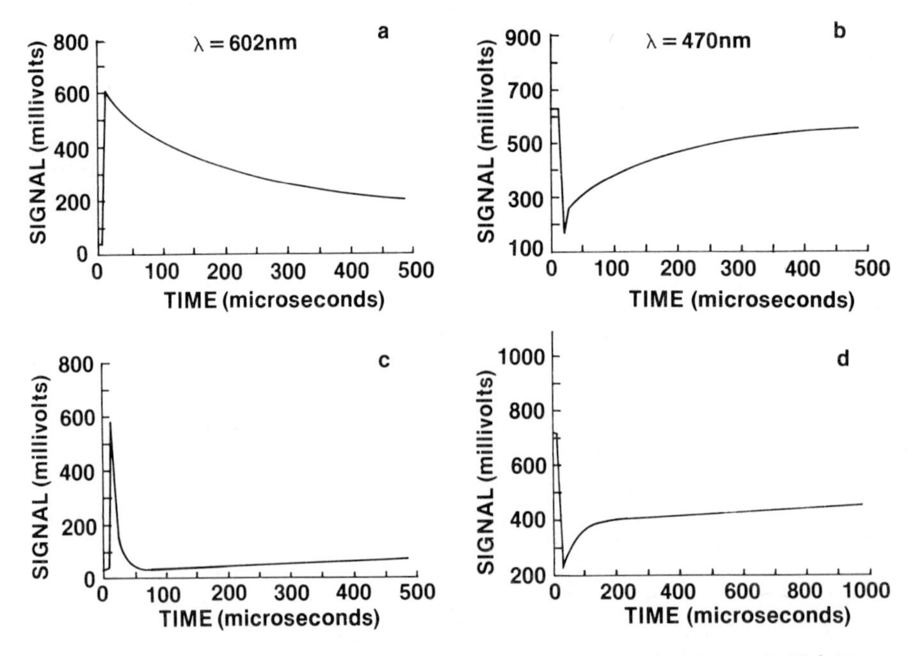

Figure 6. Oscilloscope traces obtained from the laser photolysis of 10^{-4} M $Ru(bipy)_3^{3+}$ and 2×10^{-3} M MV^{2+} in deaerated aqueous solution at pH 5. Key: a,b, without catalyst; c,d, catalyst Pt (40 mg/L) or RuO_2 (8 mg/L) loaded on 500 mg/L TiO_2.

with the oxidized sensitizer $Ru(bipy)_3^{3+}$. This finding is of crucial importance for the design of a cyclic water decomposition system where specificity and high rates of interaction are required as pointed out above.

Development of Highly Efficient Oxygen Generating Catalysts

The photodecomposition of water became feasible in a regenerative way only after oxygen generating catalysts had been developed. In 1978, we showed (36) that noble metal oxides such as PtO_2, IrO_2 and RuO_2 in macrodisperse or colloidal form are capable of mediating water oxidation by agents such as Ce^{4+}, $Ru(bipy)_3^{3+}$ and $Fe(bipy)_3^{3+}$. The development of RuO_2 has advanced rapidly since then. An impression of the improvement of the activity of RuO_2 based catalysts may be gained from the following comparison: Three years ago, in order to effect water oxidation by $Ru(bipy)_3^{3+}$

$$4Ru(bipy)_3^{3+} + 2H_2O \rightarrow 4Ru(bipy)_3^{2+} + O_2 + 4H^+ \quad (8)$$

within a time domain of several minutes we required 1g/l RuO_2 powder. Today, by using an ultrafine deposit of RuO_2 on colloidal TiO_2 particles a half lifetime of 5 to 10ms can be obtained (42) with only 4mg RuO_2/l. Thus, by decreasing the particle size of RuO_2 and stabilizing the catalyst on a suitable carrier a more than 10^6 fold increase in the catalytic activity has been achieved.

Application of combined flash photolytic and fast conductometric technique made it possible to probe the mechanistic details of the oxygen evolution reaction. Thus, $Ru(bipy)_3^{3+}$ was produced via the photoredox reaction:

$$2Ru(bipy)_3^{2+} + S_2O_8^{2-} \xrightarrow{h\nu} 2Ru(bipy)_3^{3+} + 2SO_4^{2-} \quad (9)$$

and the kinetics of oxygen production via reaction (8) were studied in the presence of a catalyst consisting of a transparent TiO_2 sol loaded with RuO_2. A comparison of the temporal behavior of the $Ru(bipy)_3^{3+}$ absorption decay and the increase in conductivity associated with water oxidation was made. Results obtained from 530nm laser photolysis experiments are illustrated in Figure 7. The temporal behavior of the $Ru(bipy)_3^{3+}$ absorption at 640nm is juxtaposed to that of the solution conductance. The lower oscillogram shows that the decrease in the absorbance occurs concomitantly with an increase in the conductivity. This indicates that hole transfer from $Ru(bipy)_3^{3+}$ to RuO_2 is immediately followed by release of protons and oxygen from water. Thus the colloidal catalyst particle couples reduction of $Ru(bipy)_3^{3+}$ to water

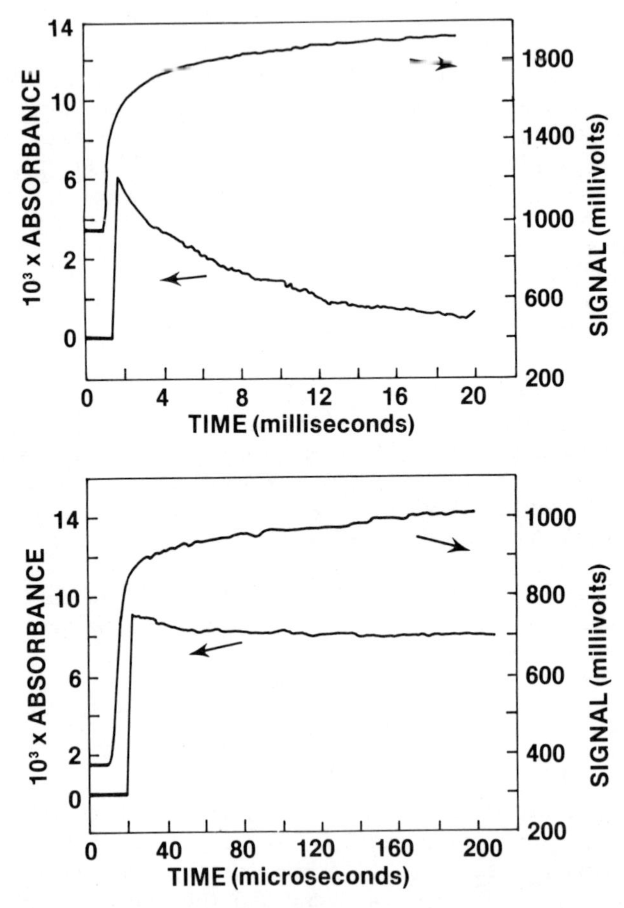

Figure 7. Laser photolysis experiments with solutions containing 10^{-4} M $Ru(bipy)_3^{2+}$ and 2×10^{-3} M $S_2O_8^{2-}$; catalyst TiO_2 (200 mg/L) loaded with RuO_2 (4 mg/L). Simultaneous observation of transient conductance and absorbance.

oxidation. This experimental observation provides a direct proof that the concept of redox catalysis, considering the RuO_2 colloid as microelectrodes, is valid for water oxidation as well as for water reduction. Moreover, The RuO_2 particles can be regarded as artificial analogues of the water splitting enzyme in photosystem II, present in chloroplasts. Both are capable of transferring four oxidation equivalents from a suitable solute to water releasing oxygen and protons. The RuO_2 mediator affords water oxidation within milliseconds at surprisingly low concentration. This proved to be extremely valuable in energy conversion systems where water is the source of photodriven uphill electron flow.

Cyclic Water Cleavage by Visible Light

Since photogenerated MV^+ and $Ru(bipy)_3^{3+}$ can be used for water reduction and oxidation respectively, it is tempting to examine a system where the two catalytic processes can take place simultaneously following photoinduced electron transfer. As was pointed out above, the RuO_2 and Pt catalysts have to be active enough to intercept the back reaction. Also, their intervention has to be specific in that MV^+ reacts selectively with the Pt particles while $Ru(bipy)_3^{2+}$ interacts with RuO_2. Cross reactions have to be avoided since they lead to short-circuitry of the back reaction.

A first successful attempt to split water photochemically this way was made by us in 1979 (43). A copolymer of maleic anhydride and styrene was used as a protective agent for the Pt sol. This is suitable to achieve selectivity since it provides functions with pronounced hydrophobicity. Of the redox products formed in the light reaction MV^+ is relatively hydrophobic and will therefore interact with the Pt. $Ru(bipy)_3^{3+}$ on the other hand is prone to interact with the hydrophilic and negatively charged RuO_2 surface. One disadvantage of this system is that the quantum yield of water splitting is small ($\sim$0.1%) and that the photo reaction stops in a closed vessel after a few hours of irradiation. One encounters here a fundamental problem which is inherent to all devices that attempt to produce photolytically H_2 and O_2 without local separation: The presence of oxygen will severely limit the quantum yield of water splitting as both depolarization of the cathodically tuned Pt particles as well as reoxidation of the reduced relay according to

$$R^- + O_2 \rightarrow O_2^- + R \qquad (10)$$

will interfere with hydrogen generation. Using computer simulation, Infelta (44) has elaborated the detailed kinetics of the processes occurring in the $Ru(bipy)_3^{2+}/MV^{2+}$ system under illumination. By taking into account the rate parameters for all relevant reactions including catalytic H_2 and O_2 production, he arrives at the conclusion that water splitting will cease once the oxygen concentration builds up in solution.

This problem has been overcome only recently through the development of bifunctional redox catalysts (45). The latter are distinguished by the fact that Pt and RuO_2 are loaded onto the same mineral carrier particle. Colloidal TiO_2 was the first material to be used as a support. It fulfills four different functions in the water splitting system (Figure 8): 1) It serves as a carrier for Pt and RuO_2 and maintains these catalysts in a highly dispersed state. 2) The TiO_2 conduction band accepts electrons from the reduced relay or the excited sensitizer. These are channeled to Pt sites where hydrogen generation occurs. As the whole TiO_2 particle is reactive the cross-section and, hence, the rate of electron capture is greatly increased with respect to systems in which polymer protected Pt particles are used as catalysts. 3) RuO_2 catalyzes oxygen production from water. 4) TiO_2 serves as an adsorbent for O_2 produced during the photolysis. Some adsorption will take place spontaneously, however, the main part is photoinduced: Electrons injected into the conduction band are used to reduce O_2 to O_2^- which is strongly attached to the TiO_2 surface. Assuming monolayer coverage 1g of TiO_2 with a surface area of $200m^2$ can adsorb ca. 50ml of O_2. Through this mechanism the amount of O_2 in solution is kept very low, which benefits greatly the efficiency of water photolysis.

A correlation has meanwhile been established between the water cleavage efficiency and the capacity for oxygen binding by TiO_2. The latter is favorably influenced by surface hydroxyl groups. Thus, a good support material is a fully hydroxylated anatase with a high surface area. Flame hydrolyzed TiO_2 has a small number of surface hydroxyl groups and, hence, a low affinity for O_2 binding. When charged with Pt and RuO_2 it shows only small catalytic activity in water decomposition systems. The binding of O_2 to the support has been demonstrated unambiguously by exposing a photolyzed solution to a high concentration of oxyanions. This provokes release of O_2 from the TiO_2 surface, which can be readily analyzed (46).

Detailed investigations have meanwhile been carried out with the TiO_2 based redox catalyst using the $Ru(bipy)_3^{2+}/MV^{2+}$ couple as a sensitizer/relay pair. Apart from the composition of the catalyst (n-doping, RuO_2 and Pt loading) the quantum yield of water splitting depends strongly on TiO_2 concentration, pH and temperature (47). Under optimum conditions, the efficiency for hydrogen production ($\phi(H_2)$) is 6% (75°C). A study of the kinetics of H_2 and O_2 generation showed that over the initial period of 10-20h irradiation time the gas released from the solution is pure

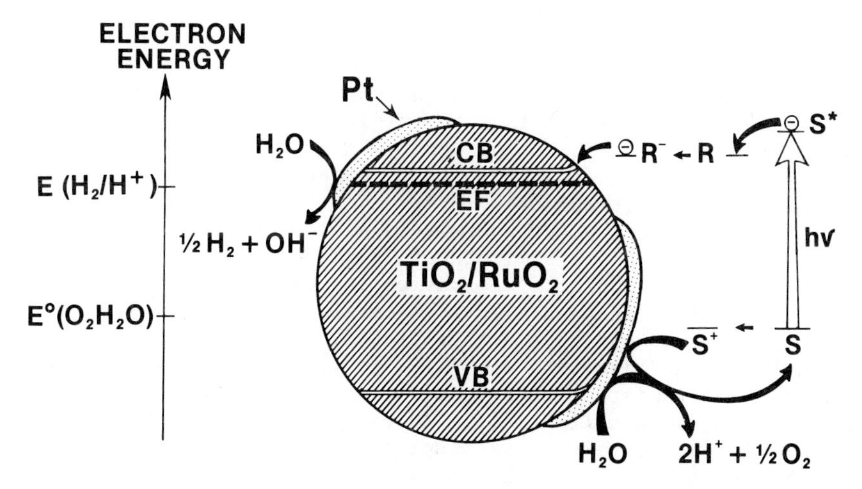

Figure 8. Schematic illustration of the intervention of a colloidal TiO_2-based bifunctional redox catalyst in the cleavage of water by visible light.

hydrogen, oxygen being retained in the solution through absorption on TiO_2. The kinetics of H_2 production from irradiating a solution containing $Ru(bipy)_3^{2+}$ as a sensitizer are illustrated in Figure 9. One observes a linear increase in the amount of hydrogen generated during the first 10 to 20h of irradiation time. Thereafter the rate of H_2 production decreases. This is attributed to the increase in oxygen concentration in the solution which lowers the quantum yield of hydrogen formation. After flushing the systems with argon the H_2 generation resumes at the initial rate. This finding is important in that it points at a way to separate hydrogen from oxygen which presents a problem for practical applications of such systems. The capacity of the carrier must be made high enough to absorb the quantity of oxygen produced from one day of solar irradiation. In such a system daylight production of hydrogen would alternate with O_2 release during the night.

Cyclic water cleavage by visible light was also achieved in electron relay free systems (48). In this case the fraction of sensitizer that is absorbed onto the particle surface is photoactive and electron injection occurs directly from its excited state into the TiO_2 conduction band. Using the surfactant ruthenium complex depicted in Figure 10, a quantum yield of 7% was obtained for the water splitting process.

A third type of water photolysis system is based on band gap excitation of colloidal semiconductors as depicted in Figure 4c. Photoinduced electron/hole separation is followed by H_2 production from conduction band electrons catalyzed by Pt. Holes in the valence band are used to generate oxygen. Previous studies have been carried out with TiO_2 or $SrTiO_3$ as support material (49-54). However, UV irradiation is required to excite these particles and efficiencies are usually small. An exception is made by the bifunctional redox catalyst which splits water with surprising efficiency under near UV illumination ($\lambda > 300nm$). Results are shown in Figure 11 which illustrates the amount of hydrogen and oxygen produced as a function of illumination time. The solution (25ml) contained 12.5mg TiO_2 loaded with 1mg Pt and 0.025mg RuO_2. Initially one obtains H_2 at a rate of 2ml/h, which corresponds to a quantum yield of ca. 30%. The rate decreases as the pressure of H_2 and O_2 builds up in solution. After flushing with argon the H_2 generation resumes at the initial rate. A key role in achieving this high efficiency is played by the RuO_2 deposit on the TiO_2 particle, which greatly facilitates the transfer of holes from the valence band of the semiconductor to the solution bulk.

This catalytic effect of RuO_2 has been exploited recently to stabilize small band gap semiconductor particles which from their absorption properties are more suitable for solar energy conversion than TiO_2. An undesirable property of these materials is that they undergo photocorrosion under illumination. Holes produced in the valence band migrate to the surface where photocorrosion occurs, i.e.,

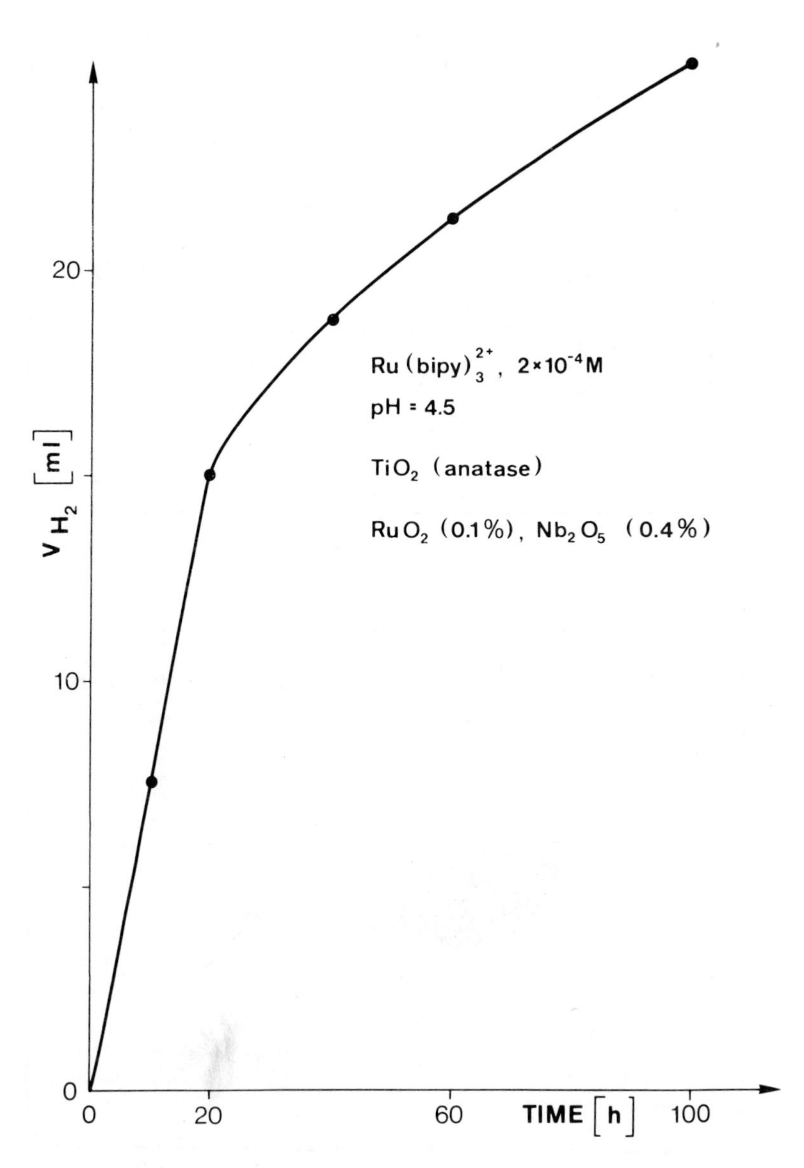

Figure 9. Cyclic water cleavage by visible light in the presence of $Ru(bipy)_3^{2+}$ as a sensitizer and bifunctional redox catalyst as an electron relay.

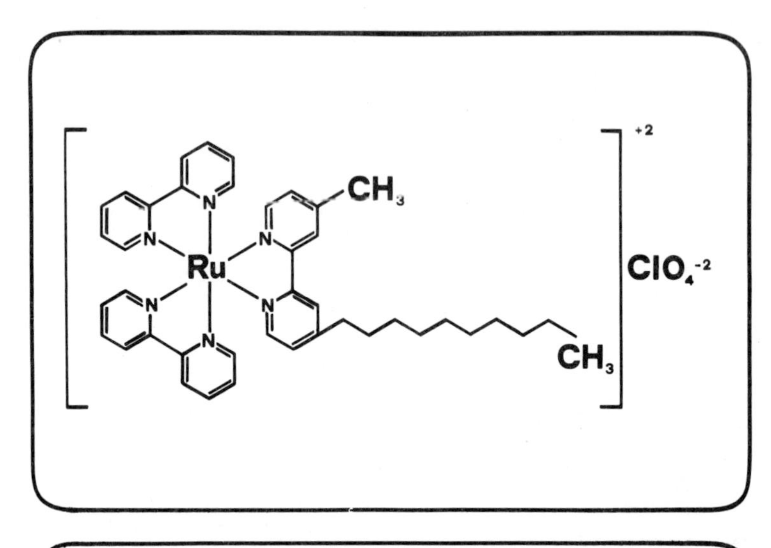

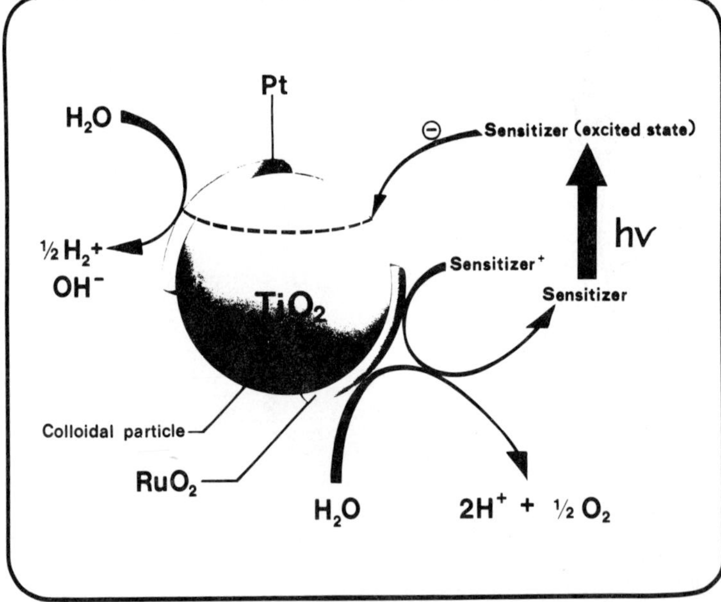

Figure 10. Processes involved in the photodecomposition of water in a relay-free system. An amphiphilic $Ru(bipy)_3^{2+}$ derivative is used as a sensitizer.

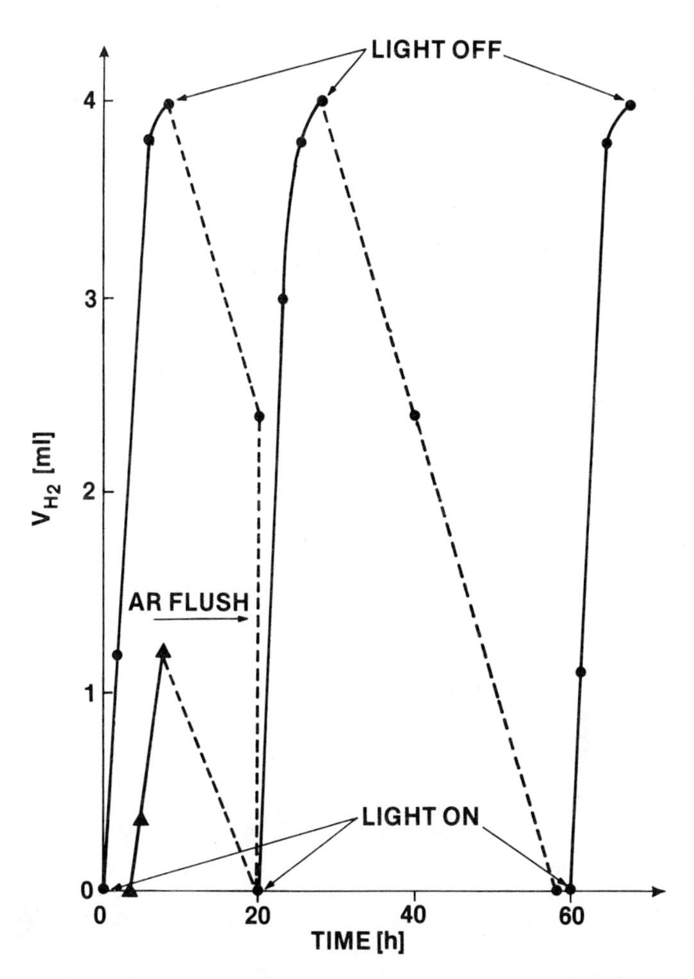

Figure 11. UV irradiation (Pyrex filtered) of the bifunctional TiO_2 (anatase) catalyst, 500 mg TiO_2/L doped with 0.4% Nb_2O_5 loaded with 0.1% RuO_2 and 40 mg Pt, pH 4.5 adjusted with HCl. Triangles indicate the volume of oxygen in mL.

$$CdS + 2^+ \rightarrow Cd^{2+} + S \quad (11)$$

Recently, we discovered (55) that loading of colloidal or macro-disperse CdS particles with an ultrafine deposit of RuO_2 prevents photodecomposition through promotion of water oxidation according to

$$4h^+(CdS) + 2H_2O \xrightarrow{(RuO_2)} O_2 + 4H^+ \quad (12)$$

Sustained water cleavage by visible light is observed when CdS particles loaded simultaneously with Pt and RuO_2 are used as photocatalyst. Again, hydrogen and oxygen are generated by conduction band electronics and valence band holes, respectively, produced by band gap excitation. Thus irradiation of a 25ml solution containing 2.5mg colloidal CdS loaded with 0.5mg Pt and 0.2mg RuO_2 yields 2.8ml H_2 and 1.4ml O_2 after 44 hours of irradiation with the visible output of a 450W Xenon lamp. The quantum yield is still significantly below that observed with the TiO_2/Pt/RuO_2 sol in the presence of a suitable sensitizer and further development is required to improve the performance of this system.

Cleavage of Hydrogen Sulfide by Visible Light

The ability of RuO_2 to promote hole transfer from the conduction band of CdS to solution species can be exploited for processes other than water oxidation. Thus, CdS suspension loaded with RuO_2 decompose very efficiently H_2S into hydrogen and sulfur under visible light irradiation (56).

$$H_2S \xrightarrow{h\nu} H_2 + S \quad (13)$$

Figure 12 illustrates the amount of H_2 produced by illuminating a suspension of 25mg CdS loaded with 0.1% RuO_2 in 25ml water of pH3. The hydrogen formation rate is constant up to more than 90% consumption of H_2S. Rates improve with pH and RuO_2 loading. For example, H_2 is produced at 9ml/h (ca. 35% quantum yield) at pH 13 with the same amount of CdS loaded by 3% RuO_2. The role of RuO_2 in this system is to enhance the rate of sulfide oxidation by holes produced in the valence band through band gap excitation. Thus electron-hole recombination is interpreted efficiently and high quantum yields are obtained. Electrons in the conduction band are used to reduce protons to hydrogen (Figure 13). Important with respect to practical application is the fact that no Pt catalyst is required to promote hydrogen formation. This is due to a cathodic shift of the flatband potential of CdS caused by absorption of sulfide ions. Thereby the driving force for water reduction to hydrogen is increased making the use of a Pt catalyst superfluous. The composition of hydrogen sulfide is an energy storing process that could become an important alternative to

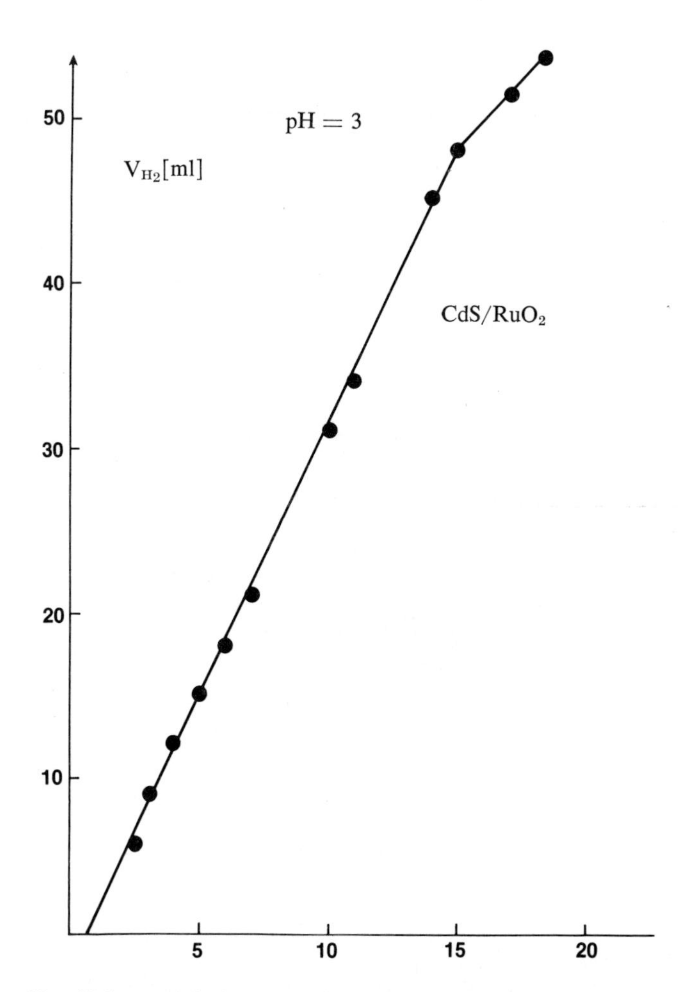

Figure 12. Volume of hydrogen evolved from irradiating in 25 mL of solution containing 0.1 M *Na_2S (pH 3) and 25 mg of CdS loaded with 0.025 mg of RuO_2 in visible light. Solution was briefly deaerated prior to illumination by flushing with Ar.*

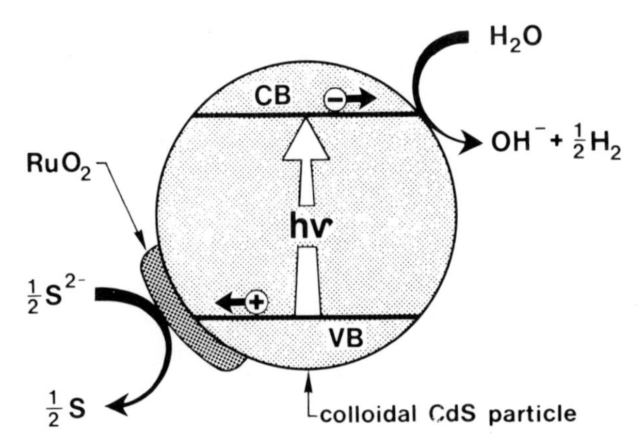

Figure 13. Schematic illustration of elementary processes involved in the CdS-sensitized decomposition of H_2S.

water cleavage as a source for hydrogen production from sunlight. H_2S is an abundant waste product in coal and petrol-related industry that could be made use of in this way.

Conclusions

Colloidal semiconductors, molecular assemblies such as micelles or vesicles and ultrafine redox catalysts provide suitable microscopic organization to accomplish the difficult task of light-induced water cleavage. Work in the future will be directed to improve the efficiency of such devices by identifying new photocatalysts and solving the problem of hydrogen/oxygen separation. Colloidal semiconductors will certainly play a primordial role in this development. Their key advantage over other functional organizations is that light-induced charge separation and catalytic events leading to fuel production can be coupled without intervention of bulk diffusion. Thus a single colloidal semiconductor particle can be treated with appropriate catalysts so that different regions function as anodes and cathodes. It appears that this wireless photoelectrolysis could be the simplest means of large scale solar energy harnessing and conversion.

Acknowledgements

The author wishes to express his deep gratitude to his collaborators as well as to his Italian colleagues Prof. E. Pelizzetti and Dr. M. Visca, whose inspired and enthusiastic efforts has made possible the success of this work. Financial assistance of the Swiss National Science Foundation, Ciba Geigy, Engelhard Industries and the United States Army Procurement Agency, Europe is also gratefully acknowledged.

LITERATURE CITED

1. Balzani, V.; Moggi, L.; Manfrin, M.F.; Boletta, F.; Gleria, M. Science 1975, 189, 852.

2. Calvin, M. Photochem. Photobiol. 1976, 23, 425.

3. Porter, G.; Archer, M.D. Interdisc. Sci. Rev. 1976, 1, 119.

4. Bolton, J. Science 1978, 202, 705.

5. Harriman, A.; Barber, j. "Photosynthesis in Relation to Model Systems"; Barber, J. (Ed.) Elsevier: Amsterdam, 1979.

6. Schumacher, E. Chimia 1978, 32, 194.

7. Grätzel, M. Ber. Bunsenges. Phys. Chem. 1980, 84, 981.

8. Paleocrassas, S.N. Solar Energy 1974, 16, 45.

9. Claesson, S. (Ed.) "Photochemical Conversion and Storage of Solar Energy"; Swedish National Energy Board Report: Stockholm, 1977.

10. Tomikwicz, M.; Fay, H. Appl. Phys. 1979, 18, 1.

11. Bolton, J.; Hall, D.O. Ann. Rev. Energy 1979, 4, 353.

12. Grätzel, M. Disc. Faraday Soc. on photoelecetochemistry, 1980, Oxford, U.K.

13. Zamaraev, K.I.; Parmon, V.N. Catal. Rev. Sci. Eng. 1980, 22, 261.

14. Bard, A.J. J. Photochem. 1979, 10, 59.

15. Kiwi, J.; Kalyanasundaram, K.; Grätzel, M. "Structure and Bonding", 1981.

16. Goodenough, J.B. Proc. Indian Acad. Sci. 1979, 88, 69.

17. Fukishima, A.; Honda, K. Nature, 1972, 238, 37.

18. Rao, K.K.; Hall, D.O. "Photosynthesis in relation to model systems", Barber, J. (Ed.) Elsevier: Amsterdam, 1979.

19. Cuendet, P.; Grätzel, M. Photobiochem. Photobiophys. 1981, 2, 93.

20. Humphry-Baker, R.; Lilie, J.; Grätzel, M. J. Am. Chem. Soc. 1981, 103, 0000.

21. Grätzel, M. Isr. J. Chem. 1979, 18, 3.

22. Grätzel, M. "Micellization and Microemulsions"; K.L. Mittal, (Ed.): Plenum Press: New York, 1977; Vol.2, P. 531.

23. Moroi, Y.; Infelta, P.P.; Grätzel, M. J. Am. Chem. Soc. 1979, 101, 573.

24. Kiwi, J.; Grätzel, M. J. Am. Chem. Soc. 1978, 100, 6314.

25. Infelta, P.P.; Grätzel, M.; Fendler, J.H. J. Am. Chem. Soc. 1980, 102, 1479.

26. Willner, I.; Ford, W.E.; Otvos, J.W.; Calvin, M. Nature 1979, 280, 823.

27. Sutin, N. J. Photochem. 1979, 10, 19.

28. Balzani, V.; Boletta, F.; Gandolfi, M.T.; Maestri, M. Topics Current Chem. 1978, 75, 1.

29. Gafney, H.D.; Adamson, A.W. J. Am. Chem. Soc. 1972, 94 8238.

30. Mauzerall, D. "The Porphyrins", Vol. V, Part c, p. 53 Academic Press, 1978.

31. Krasna, A.I. Photochem. Photobiol. 1979, 29, 267.

32. Bock, C.R.; Meyer, T.J.; Whitten, D.G. J. Am. Chem. Soc. 1974, 96, 4710

33. Koryakin, B.V.; Dzhabier, T.S.; Shilov, A.E. Dokl. Akad. Naut. SSSR 1977, 298, 620.

34. Lehn, J.-M.; Sauvage, J.-P. Nouv. J. Chim. 1977, 1, 449.

35. Chan, S.F.; Chou, M.; Creutz, C.; Matsubara, T.; Sutin, N. J. Am. Chem. Soc. 1981, 103, 369.

36. Kiwi, J.; Grätzel, M. Angew. Chem. Int. Ed. Engl. 1979, 18, 624.

37. Kalyanasundaram, K.; Micic, O.; Promauro, E.; Grätzel, M. Helv. Chim. Acta 1979, 62, 2432.

38. Kiwi, J.; Grätzel, M. J. Am. Chem. Soc. 1979, 101, 7214.

39. Grätzel, C.; Grätzel, M. J. Am. Chem. Soc. 1979, 101 7741.

40. Brugger, P.-A. Cuendet, P.; Grätzel, M. J. Am. Chem. Soc. 1981, 103, 0000

41. Duonghong, D.; Borgarello, E.; Grätzel, M. J. Am. Chem. Soc. 1981, 103, 0000

42. Humphry-Baker, R.; Lilie, J.; Grätzel, M. J. Am. Chem. Soc. 1981, 103, 0000.

43. Kalyanasundaram, K.; Grätzel, M. Angew. Chem. Int. Ed. Engl. 1979, 18, 701.

44. Infelta, P.P. to be published.

45. Kiwi, J.; Borgarello, E.; Pelizzetti, E.; Visca, M.; Grätzel, M. Angew. Chem. Int. Ed. Engl. 1980, 19, 646.

46. Borgarello, E.; Grätzel, M. to be published.

47. Borgarello, E.; Kiwi, J.; Pelizzetti, E.; Visca, M.; Grätzel, M. J. Am. Chem. Soc. 1981, 103, 0000.

48. Borgarello, E.; Kiwi, J.; Pelizzetti, E.; Visca, M.; Grätzel, M. Nature, 1981, 289, 158.

49. Sato, S.; White, J.M. J. Phys. Chem. 1981, 85, 592.

50. Schrauzer, G.N.; Guth, T.D. J. Am. Chem. Soc. 1977, 99 7189.

51. Kawai, T.; Sakata, T. Chem. Phys. Lett. 1980, 72, 87.

52. Van Damme, H.; Hall, W.K. J. Am. Chem. Soc. 1979, 101 4373.

53. Bulatov, A.V.; Khidekl, M.L. Izv. Akad Nauk SSSR Ser. Khim. 1902, 1976.

54. Lehn, J.M.; Sauvage, J. P.; Ziessel, R. Nouv. J. Chim. 1980, 4, 623

55. Kalyanasundaram, K.; Borgarello, E.; Grätzel, M. Helv. Chim. Acta 1981, 64, 362.

56. Borgarello, E.; Kalyanasundaram, K.; Grätzel, M.; Pelizzetti, E. J. Am. Chem. Soc. submitted for publication.

RECEIVED August 3, 1981.

8

The Use of Cationic Surfactants in Electrochemistry and Catalysis on Platinum

THOMAS C. FRANKLIN, MAURICE IWUNZE, and STEPHEN GIPSON

Baylor University, Chemistry Department, Waco, TX 76798

The addition of cationic surfactants, especially Hyamine 2389 (predominantly methyldodecylbenzyl trimethylammonium chloride) to aqueous systems has been shown to increase yields in electrolytic synthesis studies, to make possible voltammetric studies of various organic and inorganic compounds using platinum electrodes, and to accelerate the rate of hydrolysis of esters on platinum surfaces. The surfactant accomplishes this by solubilizing the compounds in micelles and by forming a hydrophobic film on the surface of the platinum which excludes water but allows the penetration of the reactants to the surface.

The film structure has been indicated to be similar to the structure of an inverted micelle. It is caused by adsorption of chloride ions on the platinum and the attachment of the quaternary ion by ion pairing. One can cause alternating increases and decreases in the rate of electrooxidation and catalytic esterification by the presence of monolayers, bilayers, etc.

General Use of Additives

Additives have been routinely used in corrosion (1), catalysis (2) and electrodeposition (3,4), fields in which metals interface with electrolytic solutions. Studies in these areas are part of the field of modification of metal surfaces in order to change the rates of processes occurring at the surface. In recent years there has been a good deal of work on what is known as chemical modifications of electrodes (5). While these semipermanent modifications have involved some sophisticated investigations, the additive field is largely studied by a trial and error process. The work in our laboratories has been aimed at obtaining an understanding of the role of additives in these

0097-6156/82/0177-0139$05.00/0

practical processes and in recent years this work has concentrated on cationic surfactants.

Cationic Surfactant Additives

Although anionic and cationic surfactants have always been typical additives used in the empirical studies, in recent years a more intensive look has been taken at the use of these surfactants in electrochemistry. Many of these studies have concentrated on quaternary salts (6-18) and their use in organic electrochemistry. These studies were in part stimulated by the development at Monsanto (17) and later at Phillips Petroleum (18) of electrolytic processes for the electrohydrodimerization of acrylonitrile to make adiponitrile using quaternary salts as supporting electrolytes.

The research in electroorganic chemistry has concentrated on the uses of surfactants to solubilize organic reactants and products in aqueous electrolytic solutions. In some cases large amounts of these hydrotropic salts (17,19,20) have been used to break the water structure thus increasing the solubility of the organic compounds. In other cases the surfactants have been used as emulsifying agents for organic solvents, used to dissolve the compounds (21-26). In still other cases the compounds have been solubilized in the form of micelles (27-39).

Phase Transfer and Micelle Catalysis

It has become increasingly evident that the surfactants are accomplishing more than the solubilization of the organic compounds. Certainly phase transfer catalysis would be expected to occur in the emulsion system and this has been proposed in several organic synthesis studies (21-26). The term micelle catalysis has not been used to any extent in electrochemistry. Instead terms such as ion pairing and ion bridging have been used to explain the acceleration of electrode reactions by the presence of a variety of ions in the interface between the solution and the electrode (40-42). Obviously these processes are the same king of processes one postulates in micelle catalysis.

The Surfactant Film on the Surface of the Electrode

Studies in our laboratories (43-51) have concentrated on the effects of quaternary salts on electrochemical oxidations on platinum electrodes in emulsion and micelle systems. In addition studies have been made of the effect of these surfactants on a noncatalytic process occurring at the platinum solution interface. The quaternary salt used for most of the experiments was Hyamine 2389 (predominantly methyldodecylbenzyl trimethylammonium chloride) and the aqueous solution was strongly basic. Under these conditions it was concluded (49) that the anode was covered with a layer of strongly adsorbed chloride ions. These ions were ion paired with the quaternary ion with the positive head toward the electrode and the nonpolar tail toward the aqueous solution. Thus the metal serves as a means of forming an inverted micelle (Figure 1).

Voltammetric Studies

Voltammetric methods are very useful for studying factors that influence films on the surface of metals. Figure 2 (44) shows a comparison of anodic current-voltage curves obtained with platinum electrodes in aqueous sodium hydroxide. It can be seen that in the absence of a surfactant water is oxidized at about 0.7 volts. If anionic or neutral surfactants are present there is little shift in this oxidation potential indicating that water penetrates relatively unhindered to the electrode surface. However, when Hyamine 2389 is added two things are observed. First, there is a peak at about 1.3 volts showing that the Hyamine is oxidized. Upon repeated oxidation one sees this maximum decrease in height until after 3 or 4 runs one sees a relatively flat residual current line. This behavior indicates that the Hyamine is oxidized to form a relatively firmly bound film which prevents new unoxidized Hyamine from reaching the metal surface.

In addition one sees that the oxidation potential for water is increased to about 2.0 volts. Thus this hydrophobic film has excluded water from the electrode interface to such an extent that one must apply 1.3 volts more in order to oxidize the water. This hydrophobic behavior is similar to the behavior expected in micelle systems.

Oxidation of Organic Compounds on the Filmed Surface

From a practical electrochemical viewpoint the film furnishes 1.3 volts more of oxidizing potential in which to look for the oxidation of compounds. Because of this extra potential range there are a number of compounds that give distinct oxidation waves in the presence of the Hyamine that give no wave or only indistinct waves in its absence. Figure 3 shows a voltammetric curve for the oxidation of thiourea (47). Thiourea gives no observable wave in the absence of the surfactant but gives very distinct waves in its presence. It should also be noted that the electrochemical technique furnishes a method of studying which substances are extracted by the inverted micelle into the zone of reaction.

The three effects, increased solubility, micelle catalysis, and the increased range of available oxidation potentials allows one to see normally unobserved oxidation waves for a number of compounds, a few of which are listed in Table I.

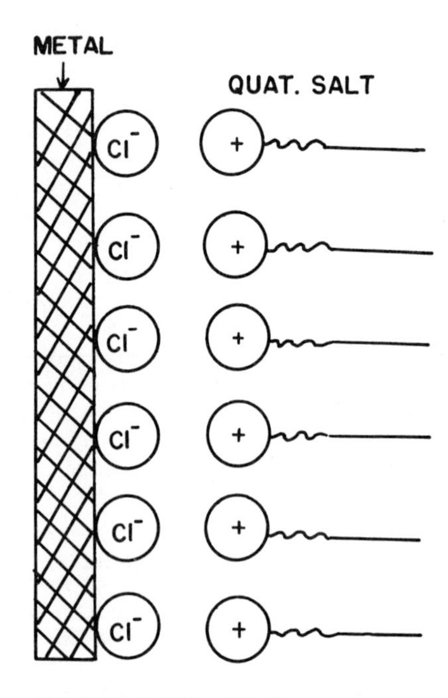

Figure 1. Simplified diagram of the monolayer film of a quaternary chloride on platinum.

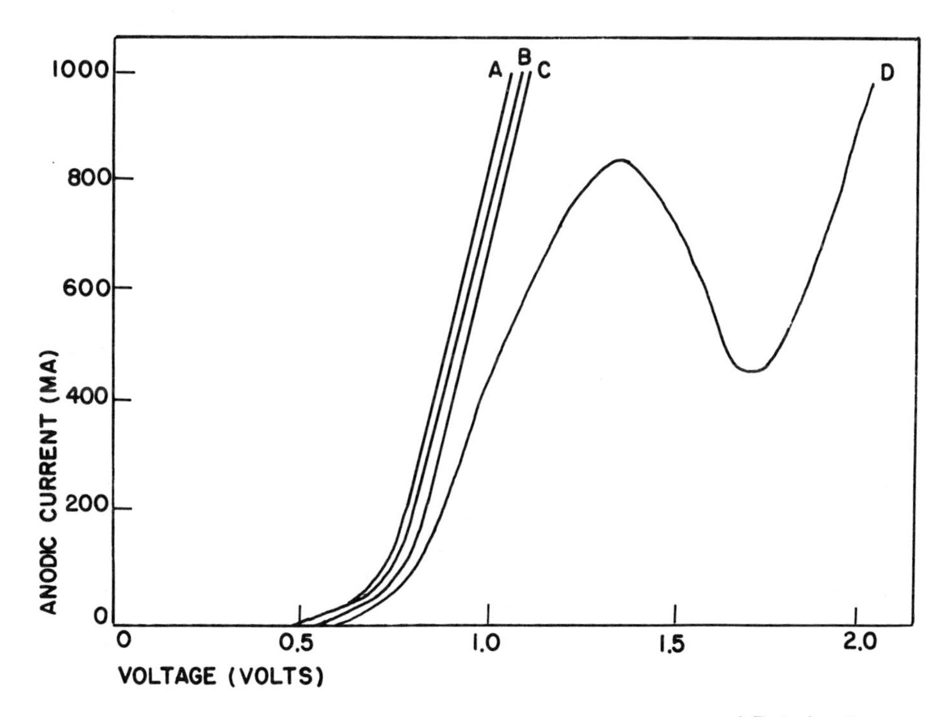

Figure 2. Effect of different surfactants on the anodic current potential (vs. SCE) curves obtained in 2 N *NaOH. Key: A, no surfactant; B, anionic surfactant; C, neutral surfactant; and D, cationic surfactant.*

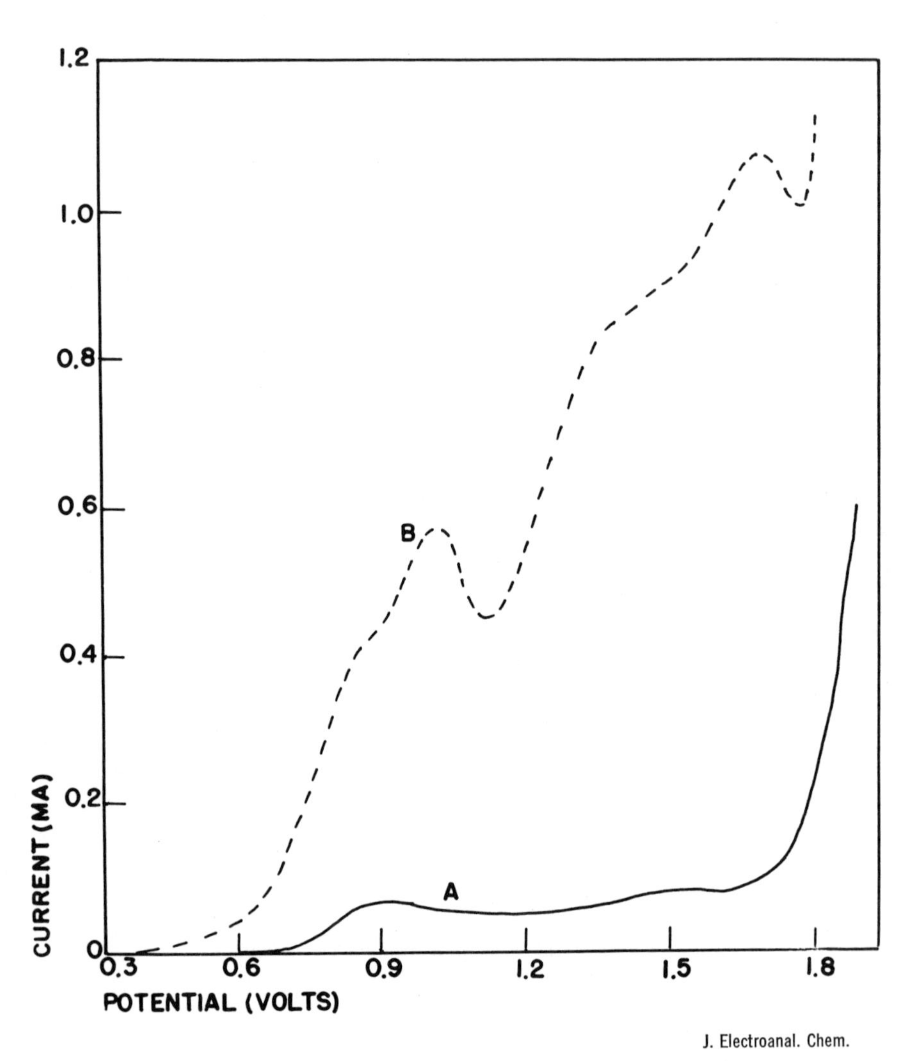

J. Electroanal. Chem.

Figure 3. Anodic current–voltage curves in aqueous sodium hydroxide. Key: A, with Hyamine 2389; and B, with Hyamine 2389 plus 8.76×10^{-3} M *thiourea.*

Table I
Observed Half-Wave Potentials of
Some Organic Compounds in
2N Aqueous Sodium Hydroxide
With and Without Hyamine Micelles

Compound	$E_{1/2}$ With Micelle	$E_{1/2}$ Sodium Hydroxide
NADH (51)	0.77, 1.15	unobserved
cysteine (51)	1.07	unobserved
2,4-dinitroaniline (48)	0.78, 0.93	unobserved
o-phenylenediamine (48)	0.42, 1.0	0.58, 1.05
p-nitrobenzoic acid (48)	0.78	unobserved
p-nitrobenzaldehyde (48)	0.77	unobserved
2,5-dichloroaniline (48)	0.85	unobserved
D-glucose (48)	0.80	unobserved
biphenyl (48)	0.86	unobserved
benzhydrol (48)	0.84	unobserved
anthracene (48)	0.90	unobserved
palmitic acid (48)	0.96	unobserved
m-toluic acid (48)	0.81	0.80(s)*
phenylsalicylate (48)	1.07	0.63(sm)
adenine (48)	1.11	0.77(s)

*Abbreviations used: s=shoulder, sm=small

Figure 4 (48) shows a comparison of oxidation potentials obtained for a variety of compounds in aqueous sodium hydroxide and in the same solution containing Hyamine, in the form of micelles and as an emulsifying agent. In aqueous solutions oxidations occur primarily around the two potentials of oxidation of platinum (The three higher oxidation potentials in the figure are in the oxygen evolution region and probably oxidation occurs by oxygen.). The mechanism of oxidation on platinum in aqueous solutions is generally accepted to be electrochemical oxidation of the platinum surface followed by a chemical reaction of the compound with the surface oxides of platinum (52).

Similarly in the emulsion system the potentials are grouped around the oxidation potential of Hyamine indicating a chemical oxidation of the compounds by the electrolytically oxidized Hyamine. However, in the micelle system the oxidations are spread over a wide range of potentials indicating direct electrochemical oxidation of the compounds. This is very similar to the results obtained in nonaqueous solutions, once more showing the hydrophobic nature of the electrode interface.

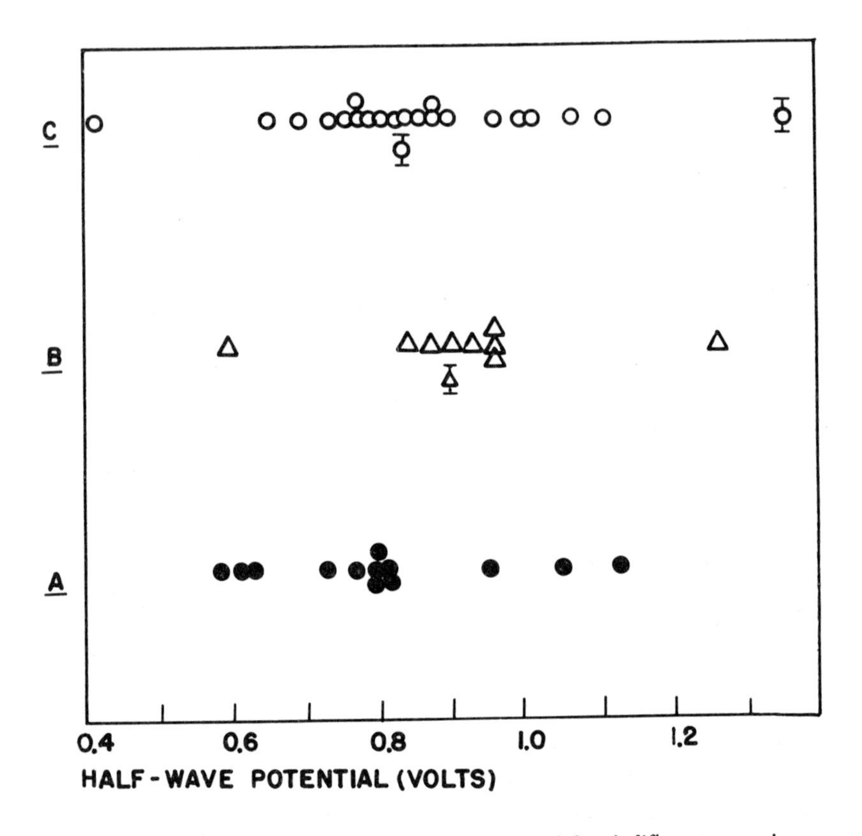

Figure 4. Comparison of anodic half-wave potentials of different organic compounds obtained in aqueous sodium hydroxide containing A, nothing; B, acetonitrile + Hyamine 2389 (emulsion); and C, Hyamine (micelle). Key: ⚲, anodic half-wave potentials of Hyamine in the micelle system; and Δ, Anodic half-wave potentials of Hyamine in the emulsion system.

Electrochemical Synthesis

One of the primary uses of modified electrodes has been in the area of electrochemical synthesis. Again the increased solubilization by micelles and emulsions has been the primary interest in using cationic surfactants. However, micelle and phase transfer catalysis and the hydrophobic nature of the electrode film has contributed to increased yields. Table II shows a comparison of yields obtained in the electrooxidation of benzhydrol in the presence of different surfactants and a comparison of the yields obtained with several other compounds with and without Hyamine 2389 (50). It can be seen that without a surfactant there is no yield in aqueous solutions. Anionic and neutral surfactants which solubilize the compound but do not film the electrode cause only small increases in yield, but the cationic film forming surfactant causes a sharp increase in yield.

Table II
The Effect of Surfactant Micelles on the Yields Obtained in Electrooxidations on Platinum Electrodes in 2N Aqueous Sodium Hydroxide*

Compound		Product	Type of Surfactant	Yield% w/surfactant
Benzyhydrol	(44)	Benzophenone	cationic	29.4
Benzyhydrol	(44)	Benzophenone	anionic	3.5
Benzyhydrol	(44)	Benzophenone	neutral	4.8
Diphenyl-acetonitrile	(45)	Dimer	cationic	34.8
Diethyl-malonate	(45)	Dimer	cationic	13.9
NADH	(51)	NAD+	cationic	48
Cysteine	(51)	Cystine	cationic	36

*In all cases negligible yields were obtained without any surfactant present.

One can observe similar effects if the same surfactants are used as emulsifying agents. Table III shows results obtained in benzene - aqueous two molar sodium hydroxide emulsions using different surfactants. Again it can be seen that the film forming cationic surfactant causes marked increases in yields. Apparently, key effect is the fact that the hydrophobic electrode film blocks the competing reaction of the electrode with water.

The hydrophobic layer also furnishes an environment which protects anodically formed free radicals. Thus it is possible to obtain appreciable yields of dimers of such compounds as diphenylacetonitrile (51) and diethylmalonate (51,52) as can be seen in Table II. In the oxidation of stilbene one can see a difference in product depending on the size of the molecule

Table III
The Effect of Surfactants on the Yields Obtained In Electrooxidations on Platinum Electrodes in Benzene-2M Aqueous Sodium Hydroxide Emulsions*

Compound	Product	Type of Surfactant	Yield %
Benzhydrol(44)	Benzophenone	Cationic	23.8
Benzhydrol(44)	Benzophenone	Anionic	2.8
Benzhydrol(44)	Benzophenone	Neutral	4.0
Diphenyl- (44) acetonitrile	Dimer	Cationic	19.7
Benzyl (44) alcohol	Benzaldehyde	Cationic	17.8
α-methyl benzyl alcohol(44)	Acetophenone	Cationic	21.3
p-methyl benzyl alcohol(44)	p-tolualdehyde	Cationic	29.9
p-nitro benzyl alcohol(44)	p-nitrobenzaldehyde	Cationic	23.3

*Without surfactant the benzhydrol had 0.4% yield and all others had 0.0% yield.

making the film. With the smaller tetraethylammonium chloride where water can attack the radical one obtains predominantly the ketone, but when Hyamine 2389 is used the product is an oil.

It should be pointed out that these are not good synthesis methods. It is difficult to separate the product from the surfactant. The yields listed in Table II are the isolated yields. Undoubtedly the true yields are higher. Coulometric studies indicate that in several cases the yields approach 100%.

Heterogeneous Catalysis

Micelle catalysis of such reactions as the hydrolysis of ethyl benzoate have been extensively studied (53,54). Although platinum does not normally catalyze the hydrolysis, if one inserts a piece of platinum into a solution in which the micelle catalyzed reaction is occurring the rate is accelerated (50). The increase in rate is linearly proportional to the area of the platinum (Figure 5). If one varies the concentration of the surfactant one sees a periodic rise and fall in the extra catalysis caused by the platinum. The increase in catalysis by platinum rises to about 4.0×10^{-6} M/sec at 22 mM Hyamine, decreases to about 0 at 45 mM, increases to 3.8 M/sec at 95 mM. A logical explanation of this data can be obtained from the structure of the film. Figure 6 shows a simplified model of the filmed electrode with multilayers present. As one adds small amounts of surfactant one forms an inverted micelle and obtains the extra catalysis of the inverted micelle causing a rise in the rate. As more surfactant is added the normal micelle starts to form on the surface and the rate drops back toward the catalysis of the normal micelle. This process is repeated through the second and third layers (50).

That one is looking at catalysis by an adsorbed film can further be shown by potentiostatting the metal at various potentials (50). At +1.4 volts where the chloride is more strongly adsorbed the reaction rate is 11.8×10^{-6} M/sec. It decreases as the potential decreases reaching a minimum of about 7.8×10^{-6} M/sec around the zero point of charge and then begins to increase again reaching 11.8×10^{-6} M/sec again around -0.6 V. The increase in the negative potential region is probably due to direct adsorption of the quaternary ion.

Iodide Oxidation

Because inorganic systems do not generally need the solubilizing ability of the surfactants studies of inorganic systems have been limited to such fields as the role of additives in electroplating. There is however an interest in studying simple inorganic ions to determine what type of substance will penetrate the film. Most of the work in our laboratory has concentrated on the iodide ion. Iodide gives no oxidation wave in aqueous 2N

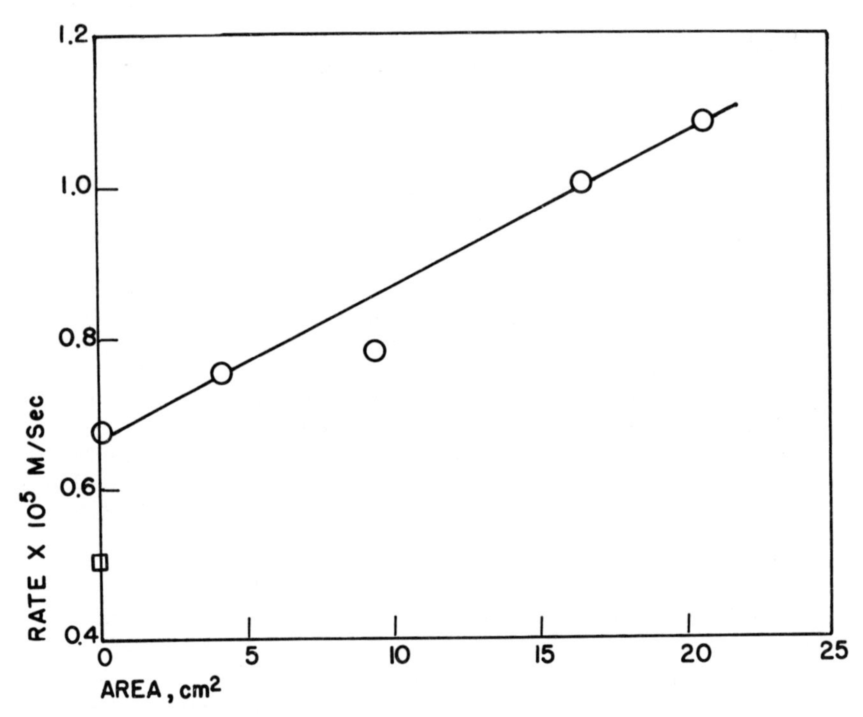

Figure 5. The effect of added platinum metal on the hydrolysis of ethyl benzoate in the presence of Hyamine 2389 (4.87 × 10⁻² M). Key: □, rate in the absence of platinum or surfactant, and ○, rate in the presence of platinum and surfactant.

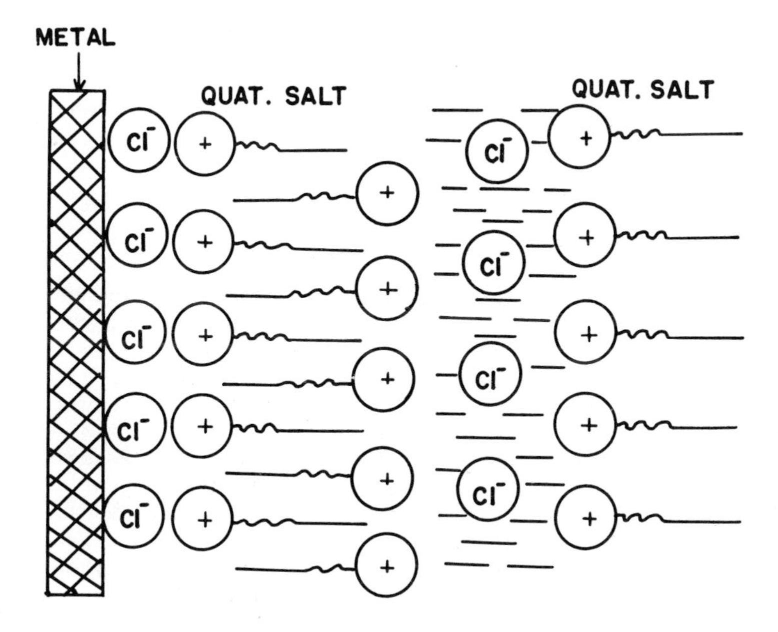

Figure 6. Simplified schematic of multilayer films formed on platinum by quaternary chlorides.

sodium hydroxide but when Hyamine is added to the solution an oxidation wave showing two maxima develops. The height of these waves are linearly proportional to the concentration of the iodide (49).

The product of the oxidation of iodide on the film is iodate. Table IV shows that the yield of iodate increases when Hyamine is added to the solution and that current efficiency increases. The latter fact is caused by the decrease in the side reaction, the electrolysis of water.

Table IV
The Effect of Hyamine on the Electrooxidation of Iodide in Aqueous 2N Sodium Hydroxide (49)

	NaOH(aq.)	NaOH(aq.) + Hyamine
%Yield of Iodate	2.6	24.0
%Current Efficiency	0.28	30.4

Changes in concentration of Hyamine causes the same periodic variation in the rate of oxidation of iodide as was observed in the hydrolysis of ethyl benzoate. This again shows the influence of multilayers of surfactant on the rate of the electrode reaction. It further indicates that oxidation of iodide occurs at a distance from the electrode, as long as it is in the film (40).

Most other ions do not penetrate the film. For example Figure 7 shows that ferrocyanide, without Hyamine present gives a simple oxidation wave. With Hyamine present the normal wave disappears indicating that the ferrocyanide cannot penetrate to the electrode. However, there is a sharp peak at the potential at which the Hyamine normally begins to oxidize. The residual then decreases to a very low current. This film which is quite impervious to any of the compounds that are normally oxidized on the Hyamine film (49) probably consists of a Hyamine-ferrocyanide insulating layer.

Acknowledgment

We wish to thank The Robert A. Welch Foundation of Houston for their support in all of these studies.

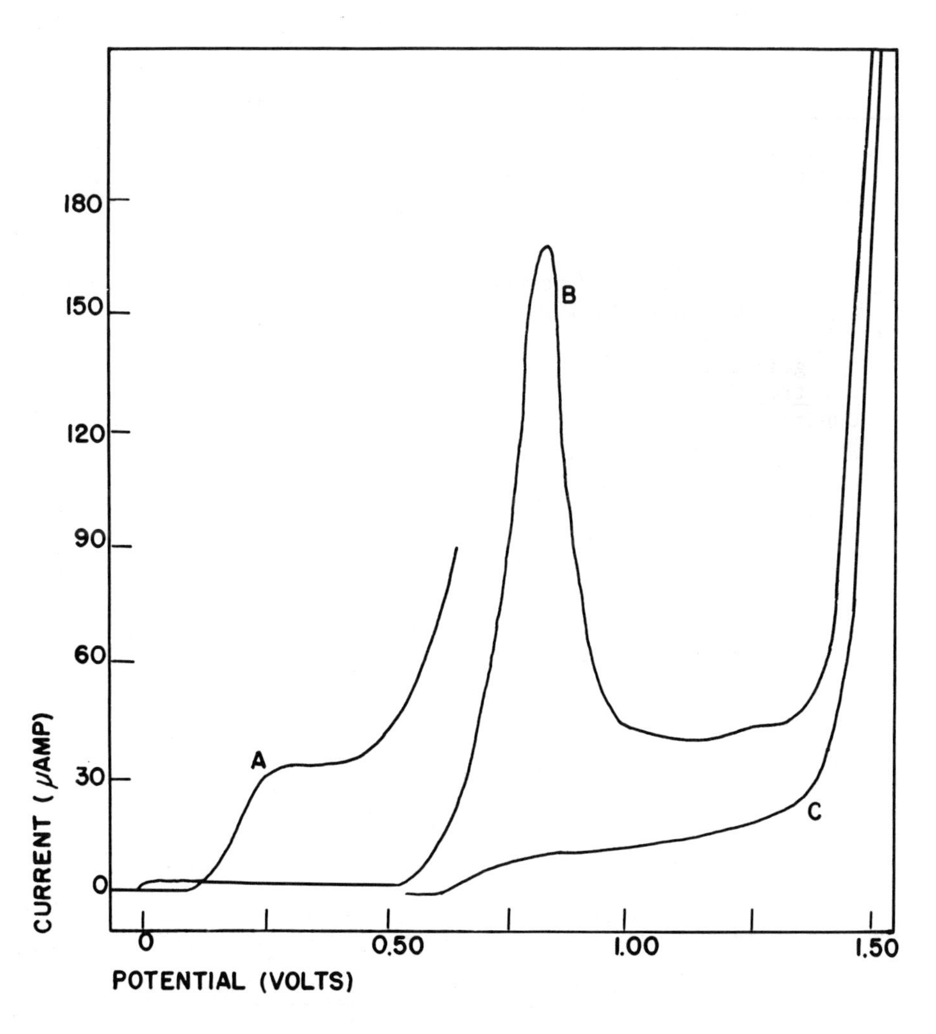

Figure 7. Current–voltage curve for oxidation of ferrocyanide in 2 N *NaOH. Key: A, 3.3* $\times$ *10*$^{-4}$ M *Fe(CN)*$_6^{-4}$*; B, 3.3* $\times$ *10*$^{-4}$ M *Fe(CN)*$_6^{-4}$ + *Hyamine (1st run); and C, same as B (4th run).*

Literature Cited

1. Ranney, M. W. "Corrosion Inhibitors. Manufacture and Technology," Noyes Data Corporation: Park Ridge, New Jersey, 1976.
2. Ashmore, P. G. "Catalysis and Inhibition of Chemical Reactions," Butterworths: London. 1963.
3. Brown, H. Plating 1968, 55, 1047-55.
4. Kodina, I. P.; Loshkarev, M. A.; Loshkarev, Y. M. Electrokhimiya 1977, 13, 715-20.
5. Murray, R. W. Accounts of Chem. Research 1980, 13, 135-41.
6. Horner, L. "Onium Compound," in Organic Electrochemistry, ed. Baizer, M. M.; Marcel Dekker: New York, NY, 1973, 429-44.
7. El-Samahy, A. A.; Ghoneina, M. M.; Issa, I. M.; Tharwat, M. Electrochim. Acta 1972, 17, 1251-9.
8. Mairanovshii, S. G.; Proskurovskaya, I. V.; Rubinskaya, T.Ya. Elektrokhimiya 1974, 10, 1502-6.
9. Mairanovskii, S. G. Elektrokhimiya 1969, 5, 757-9.
10. Chargelishvili, V. A.; Dzhaparidze, D. I.; Shavgulidze, V.V. Elektrokhimiya 1974, 10, 1414-17.
11. Gunderson, N.; Jacobsen, E. J. Electroanal. Chem. Interfacial Electrochem. 1969, 20, 13-22.
12. Nyberg, K. J. Chem. Soc. 1969, 13, 774-5.
13. Martigny, P.; Simonet, J. J. Electroanal. Chem. Interfacial Electrochem. 1979, 101, 275-9.
14. Fischer, Hellmuth J. Electroanal. Chem. Interfacial Electrochem. 1975, 62, 163-78.
15. Simonet, J.; Lund, H. J. Electroanal. Chem. Interfacial Electrochem. 1977, 75, 719-30.
16. Piccardi, G. J. Electroanal. Chem. Interfacial Electrochem. 1977, 84, 365-72.
17. Baizer, M. M. J. Electrochem. Soc. 1964, 111, 215.
18. Childs, W. V.; Walter H. C. A I Ch E Symp. Ser. 1979, 75, 19-25.
19. Brockmann, C. J.; McKee, R. H. Trans. Electrochem. Soc. 1932, 62,203.
20. Gerapostolou, B. G.; McKee, R. H. Trans. Electrochem. Soc. 1935, 68, 329.
21. Eberson, L.; Helgee, B. B. Chem. Scri. 1974, 5, 47.
22. Eberson, L.; Helgee, B. Acta Chem. Scand. 1975, B29, 451.
23. Eberson, L.; Helgee, B. Acta Chem. Scand. 1978, B32, 157.
24. Hayano, S.; Shinozuka, N. Bull. Chem. Soc. Japan 1969, 42, 1469.
25. Eberson, L.; Helgee, B. Acta Chem. Scand.1978, B32, 157.
26. Eberson, L.; Helgee, B. Acta Chem. Scand. 1978, B31, 813.
27. Pletcher, D. Tomov, N. J. Appl. Electrochem. 1977, 7, 501.
28. Hayano, S.; Shinozuka, N. Bull. Chem. Soc. Japan 1970, 43, 2083.
29. Hayano, S.; Shinozuka, N. Bull. Chem. Soc. Japan 1971, 44, 1503.

30. Day, R. A. Jr.; Underwood, A. L.; Westmoreland, P. G. Anal. Chem. 1972, 44, 737.
31. Erabi, T.; Huira, H.; Tanaka, M. Bull. Chem. Soc. Japan 1975, 48, 1354.
32. Franklin, T. C.; Sidarous, L. Chem. Comm. 1975, 741.
33. Proske, G. E. O. Anal. Chem. 1952, 24, 1834.
34. Hayaon, S.; Shinozuka, N. Bull. Chem. Soc. Japan 1969,42,1469
35. Day, R. A. Jr.; Underwood, A. L.; Westmoreland, P. G. Anal. Chem. 1972, 44, 737.
36. Erabi, T.; Hiura, H.; Tanaka, M. Bull. Chem. Soc. Japan 1975, 48, 1354.
37. Leh, Peter; Kuwana, T. J. Electrochm Soc. 1976, 123, 1334-9.
38. Smith, J. D. B.; Phillips, D. C.; Davies, D. H. J. Polym. Sci., Polym. Chem. Ed. 1977, 15, 1555-62.
39. Noel, M.; Anantharaman; Udupa, H. V. K. Electrochim. Acta 1980, 25, 1083.
40. Gerischer, H. Z. Elektrochem. 1960, 54, 366.
41. de Levie, R. J. Electrochem. Soc. 1971, 118, 185c.
42. Abubacker, K. M.; Malik, W. U. J. Indian Chem. Soc. 1959, 36, 463.
43. Franklin, T. C.; Sidarous, L. Chem. Comm. 1975, 741.
44. Franklin, T. C.; Sidarous, L. J. Electrochem. Soc. 1976, 124, 65-69.
45. Franklin, T. C.; Honda, T. Micellization, Solubilization, and Microemulsions, edited by Mittal, 1977, 2, 617-626.
46. Franklin, T. C.; Honda, T. Electrochimica Acta 1978, 23, 439-444.
47. Franklin, T. C.; Iwunze, M. J of Electroanalytical Chemistry, 1980, 108, 97-106.
48. Franklin, T. C.; Iwunze, M. Analytical Chemistry 1980, 52, 973-976.
49. Franklin, T. C.; Gibson, S. Article in preparation.
50. Franklin, T. C.; Iwunze, M. Article submitted.
51. Franklin, T. C.; Iwunze, M. Paper presented at Spring, 1980 Electrochemical Society Meeting in St. Louis.
52. Liang, C.; Franklin, T. C. Electrochim. Acta 1964, 9, 517.
53. Bunton, C. A. Pure and Applied Chem. 1977, 49, 969.
54. Fendler, J. H. Accts. Chem. Res. 1976, 9, 153.

RECEIVED August 12, 1981.

9

Reactions of Long-Chain Acidato Complexes of Transition Metals in Micelles and Microemulsions

G. MATTNEY COLE, JR.

University of Georgia, Department of Chemistry, Athens, GA 30602

Complexes of Co(III) and Rh(II) have been prepared having the general formulas $RM(NH_3)_5$ and $R'M-(NH_3)_4$ where R = $CH_3(CH_2)_6CO_2^-$, $CH_3(CH_2)_4CO_2^-$, $CH_3(CH_2)_2CO_2^-$, $CH_3CH_2CO_2^-$ and $CH_3CO_2^-$, and R' = a substituted phenanthroline. We have examined the rates for aquation reactions and condensation reactions with several β-diketones in both micelles and water in oil microemulsions. Overall rates in micelles are generally slower than in aqueous solution for the shorter chain acids. Longer chain acid complexes show rate enhancement in micellar solutions but tend to undergo a base hydrolysis side reaction.

Inorganic chemists investigating reaction mechanisms are often at a disadvantage relative to organic chemists. Many inorganic reactions of interest occur too fast to be accessible by normal techniques. Furthermore, one generally cannot tailor inorganic complexes to obtain mechanistic parameters.

However, it is well known that micelle and microemulsion solutions can have a profound effect on reaction rates for organic(1) and inorganic(2) reactions. In this work we report mechanistic data from microemulsion solutions that would otherwise be inaccessible.

The results of these investigations have yielded, in addition to reaction mechanisms, useful data on the structure of the microemulsion solutions, and on the properties of the complexes themselves.

Acidato Complexes

The cobalt acidato complexes are readily prepared from the pentammineaquacobalt(III) perchlorate, $Co(NH_3)_5H_2O\ (ClO_4)_3$. As reported elsewhere(3), the aqua complex is added to a 10-fold

0097-6156/82/0177-0157$05.00/0

excess of a 1:4 mixture of the free organic acid and its sodium salt and heated at 60°C overnight. The reaction mixture is rotovaped and recrystallized from an ethanol-water solution. Alternatively, acidato complexes can be prepared in better yields from the pentamminecarbonato-cobalt(III) nitrate, $Co(NH_3)_5CO_3 NO_3$.

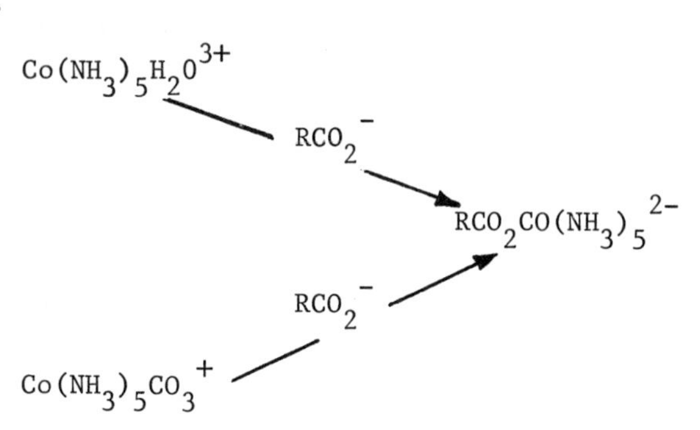

$R = CH_3, CH_3CH_2, CH_3(CH_2)_4, CH_3(CH_2)_6, CH_3(CH_2)_8$

The hexanato, octanato, and decanato complexes are unique, interesting complexes, and, to the best of our knowledge, have not been previously reported. These complexes are surface-active in aqueous solution. They form stable solutions, presumably microemulsions, with alcohols such as 2-propanol, cyclohexanol, benzylalcohol and toluene. They also form stable microemulsions with nonionic detergents, such as Tween 20 and Triton-X, and with cationic detergents, such as hexadecyltrimethylammonium bromide (CTAB). With anionic detergents, the acidato complexes form insoluble precipitates at detergent concentrations near the cmc, but dissolve as the detergent concentration is increased.

Hydrolysis of Acidato Complexes

We have investigated the kinetics of base hydrolysis reactions of the cobalt acidate complexes in aqueous solution and in several microemulsion solutions in which detergent concentrations are at least twice the respective cmc. The results are complicated by the onset of a slow secondary reaction which is presumably formation of insoluble, polymeric hydroxo, or hydrated hydroxo compounds.

$$RCO_2Co(NH_3)_5^{2+} + OH^-/HOH \longrightarrow HOCo(NH_3)_5^{2+} + RCO_2^-$$

$$\frac{Co(OH)}{(?)}\ x$$

Nevertheless, we were able to follow the reaction by spectrophotometric means over about two half-lives, particularly for the faster reactions in CTAB solution. The observed relative rates are shown in Table I. Figure 1 shows the observed rate constants plotted as a function of chain length.

Table I.

RELATIVE RATES OF BASE HYDROLYSIS
OF $RCO_2Co(NH_3)_5^{2+}$ IN VARIOUS SOLUTIONS

R	AQUEOUS SOLUTION	CTAB	TWEEN-20 TRITON-X	AQUEOUS +25% IPA	SDS
C_6	$3.2x10^{-3}$	$5.8x10^{1}$	$9.1x10^{-4}$	N/A	Very Slow
C_8	0.52	$4.3x10^{2}$	$6.3x10^{-3}$	$8.4x10^{-3}$	"
C_{10}	1	$7.5x10^{2}$	$1.2x10^{-2}$	$1.5x10^{-2}$	"

$$\text{Relative Rate} = \frac{k_{OBS}}{k_{OBS}(C_{10}, \text{ aqueous})}$$

The reaction is found to be first-order in complex and approximately first-order in hydroxide. The observed rate is independent of other added nucleophiles, such as thiocyanate ion, SCN^-. Substantial rate enhancement is observed in hexadecyltrimethyl-ammonium bromide, but there is not evidence of bromide participation; that is, there is no evidence of formation of $BrCo(NH_3)_5^{2+}$.

Discussion

The orientation of the acidato complex is shown in Figure 2. The complex resides at the interface. Since hydrogen-bonding between hydroxide ion and the ammine hydrogen is undoubtedly important, the initial step in the reaction is thought to be abstraction of proton by hydroxide, forming a amido complex (Figure 2). The amido complex can dissociate through one of two possible five-coordinate intermediates; either tbp complex or a square pyramidal complex (Figure 3). Since the observed rate increases as a function chain length, the square pyramid seems the more reasonable intermediate.

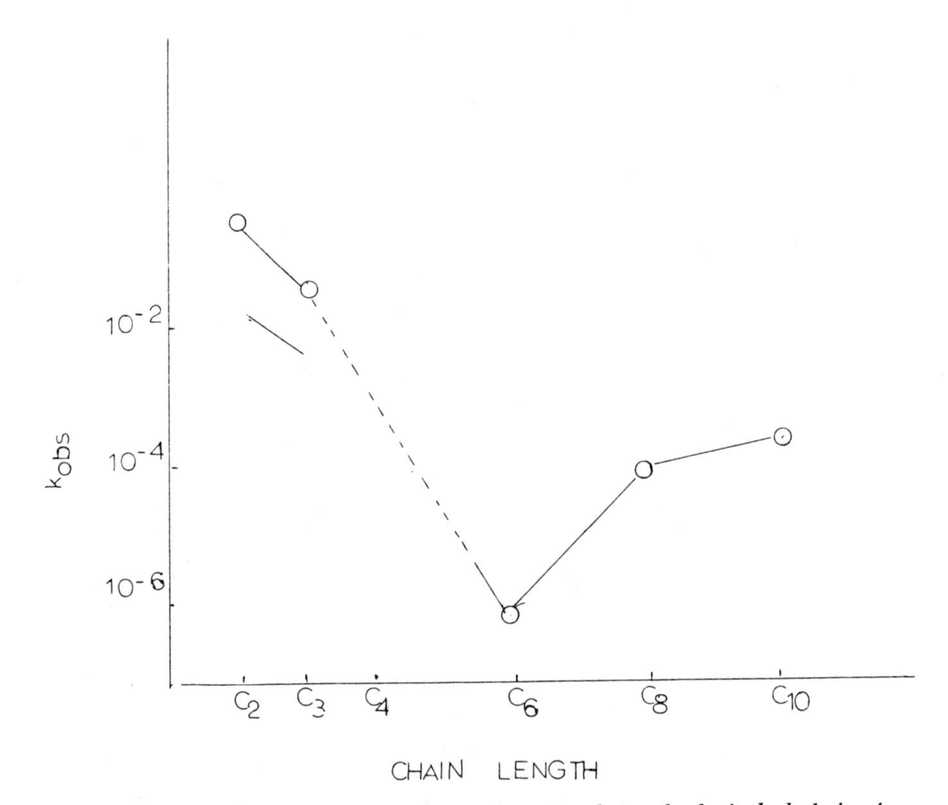

Figure 1. Observed rate constants vs. chain length for the basic hydrolysis of $RCO_2Co(NH_3)_5^{2+}$ in an aqueous hexadecyltrimethylammonium bromide (CTAB) solution. Key: ○, this work.

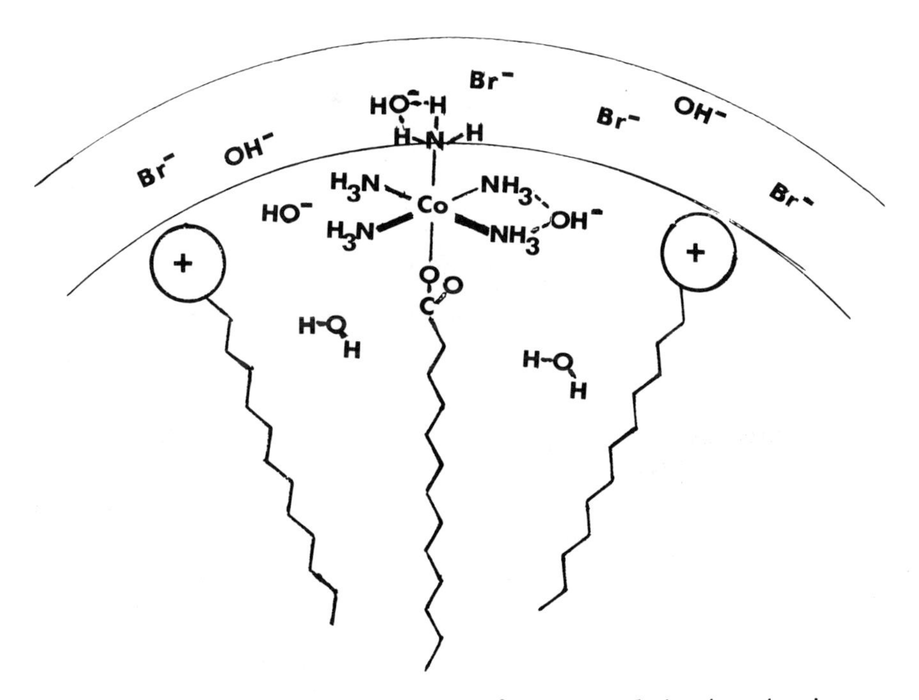

Figure 2. Orientation of the $RCO_2Co(NH_3)_5^{2+}$ complex at the interface of a microemulsion containing hexadecyltrimethylammonium bromide (CTAB).

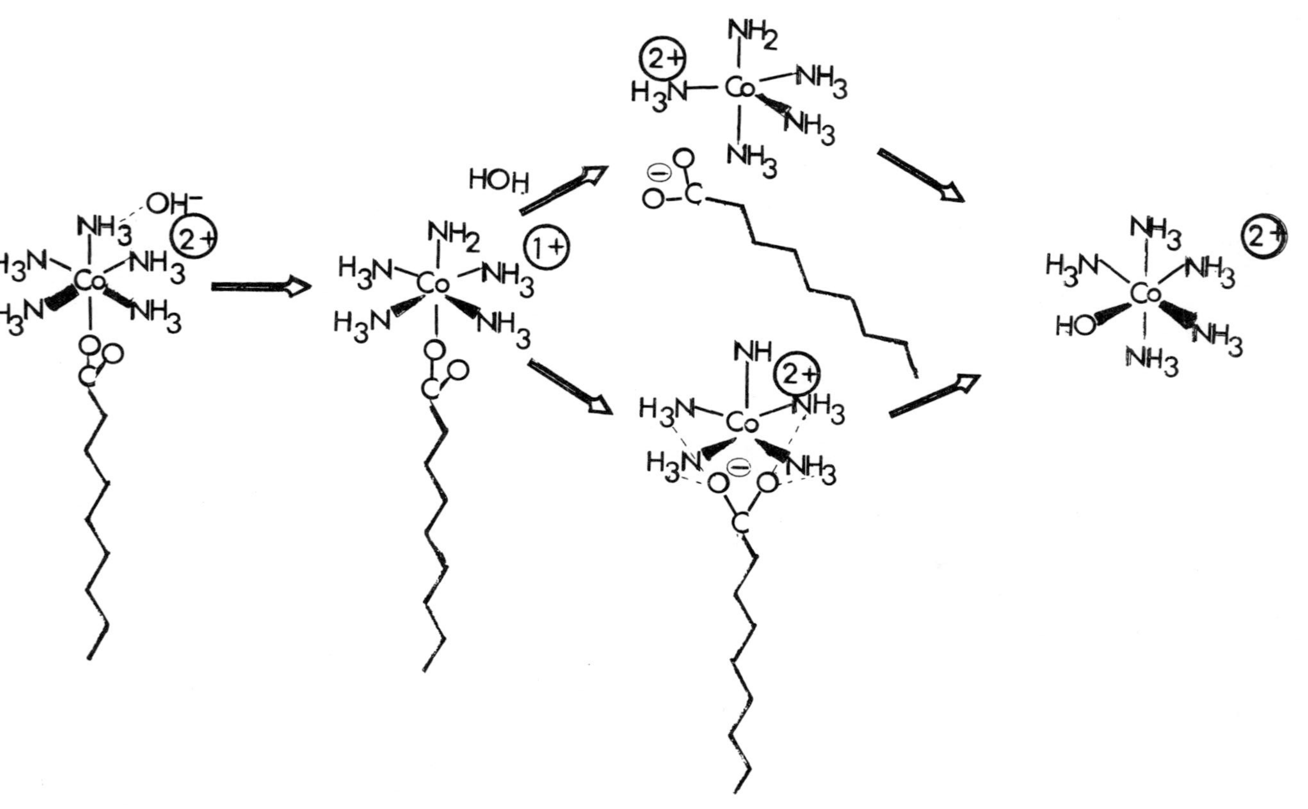

Figure 3. Postulated mechanism for the basic hydrolysis of $RCO_2Co(NH_3)_5^{2+}$ in an aqueous CTAB solution.

Literature Cited

1. Fendler, J. H. and Fendler, E. J. "Catalysis in Micellar and Microemulsion Systems", Academic Press, New York, 1975.
2. Letts, K, and Mackay, R.A., Inorg. Chem., 1975, 14, 2990.
3. Hearn, B. and Cole, G. M., (to be submitted to Inorg. Chem.).

RECEIVED September 1, 1981.

10

Chemical Reactions in Water-in-Oil Microemulsions

ALVORO GONZALEZ and JOHN MURPHY
University of Georgia, Department of Chemistry, Athens, GA 30602
SMITH L. HOLT
Oklahoma State University, Department of Chemistry, Stillwater, OK 74078

We have carried out a variety of chemical reactions in microemulsions. These include the metalation of meso-tetraphenylporphine, the base hydrolysis of long chain esters, the syntheses of macrocyclic lactones and the catalytic formation of ketones. In all instances there is a clear demonstration of the effect of microemulsification on reaction rate and pathway.

Microemulsions as media for chemical reactions have only recently received close scrutiny. This neglect arose, in part, because of the limited number of carefully characterized microemulsion systems and, in part, because strong sentiment existed that microemulsions were in actuality merely swollen micelles. Current thinking suggests that there is indeed a difference between micellar solutions and microemulsion media, and that difference is such that reaction rates and pathway need not be similar in the two media(1). (For a current review of the literature on microemulsions see Ref. 1.)

Micelles can exist as two component systems consisting of an amphiphile dissolved in either water or a hydrocarbon. When amphiphile sufficient to exceed the critical micelle concentration is dissolved in water, a "normal" micelle is formed, Figure 1a, i.e. the hydrophobic tails of the surfactant are directed inward while the polar head groups are in contact with the aqueous external phase. If a hydrocarbon is the bulk phase, the hydrophobic tails of the amphiphile will be directed outward, creating an "inverse" micelle, Figure 1b. Water added to an inverted micellar solution, is not distributed evenly throughout the hydrocarbon continum, but is found associated with the amphiphilic head groups. This is termed a "swollen inverse" micelle, Figure 1c. The volume of water which can be taken up and stabilized in these swollen inverse micelles is limited, usually only a small fraction of a mole percent of the total liquid present in the system.

0097-6156/82/0177-0165$05.00/0

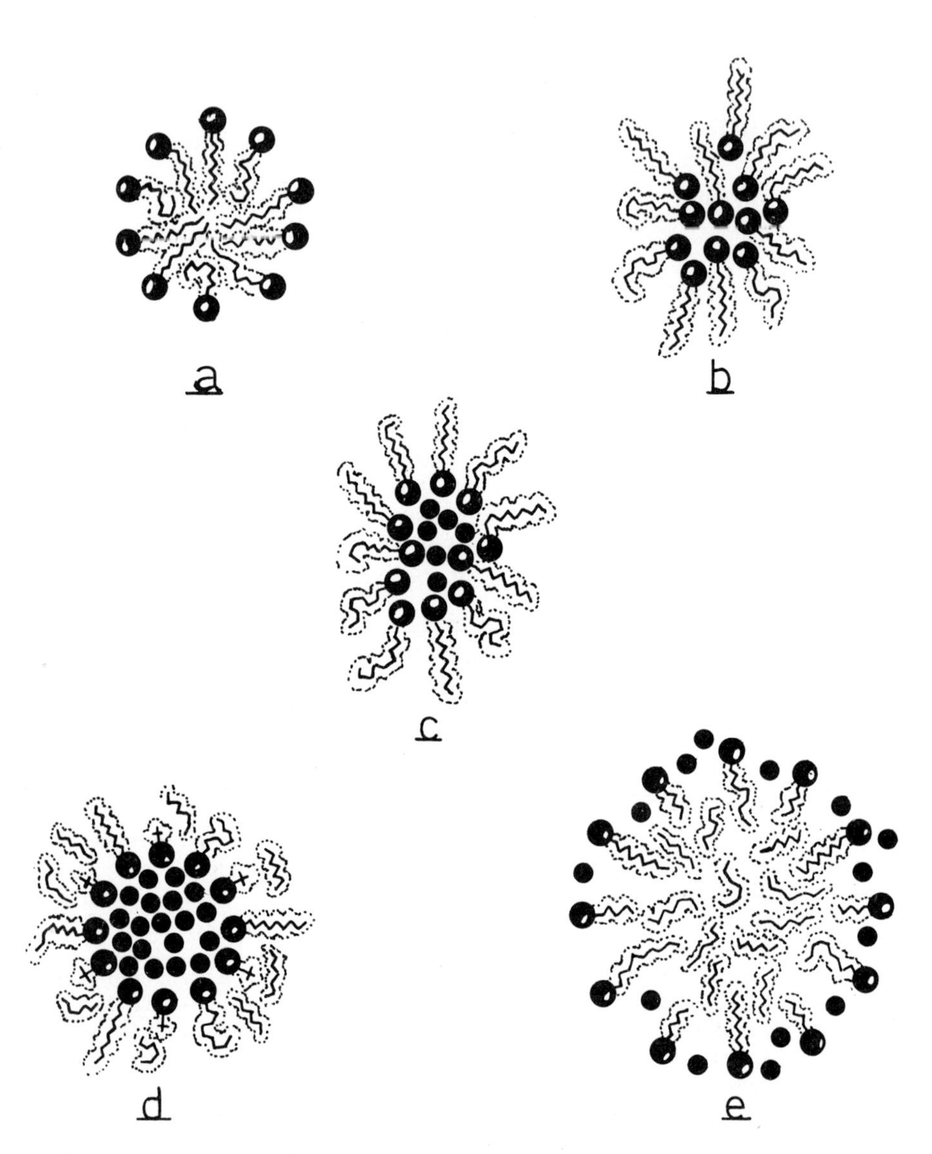

Figure 1. Some organized assemblies: a, "normal" micelle; b, "inverse" micelle; c, "swollen inverse" micelle; d, water-in-oil microemulsion; and e, oil-in-water microemulsion. Key: ●〰, surfactant; ●, water; ●+, 2-propanol; and 〰, hexane.

Microemulsions are related to micelles(1). The most common, the four component microemulsions, are constructed from a hydrocarbon, a surfactant, a short chain alcohol (cosurfactant) and water. When the hydrocarbon component present is significantly larger than the water component the microemulsion is generally a water-in-oil (w/o) microemulsion, Figure 1d. This designation arises by virtue of the fact that the water is present in the form of spheres, invisible to the naked eye (250Å to 1000Å in diameter), dispersed throughout the hydrocarbon continuum. The surfactant and cosurfactant stabilize these water-rich droplets and help render them thermodynamically stable. These systems are optically transparent and can contain up to $0.3X_{H_2O}$(3,4). As a consequence of the large mole fraction of water present, w/o microemulsions display a much greater ability to solubilize polar reactants than swollen inverted micelles. Similarly, oil-in-water microemulsions, o/w, Figure 1e. show an enhanced propensity to dissolve non-polar reactants when compared to normal micelles.

More recently it has been demonstrated that microemulsions can be formed using only water, hydrocarbon and 2-propanol, omitting the addition of a conventional surfactant. These "detergentless" microemulsions have been constructed using either hexane(2) or toluene(3) as the hydrocarbon phase. The properties of these systems have been shown to be similar to those which contain long chain amphiphiles(2-5).

The ability of micellar solutions and microemulsions to dissolve and compartmentalize both polar and non-polar reactants has a significant effect on chemical reactivity. An idealized representation of a typical micelle catalyzed reaction is depicted in Figure 2. Here the non-polar reactant is solubilized within the micelle while the ionic reactant is at the surface. The polar head groups of the surfactants generate a charge at the micelle surface which serves to attract an oppositely charged water soluble reactant increasing the concentration of that reactant near the micelle. The result is an enhanced reaction rate.

Microemulsions work much in the same way; in an o/w microemulsion, the non-polar reactant is dissolved in the oil droplet, with the polar reactant in the water continuum. Chemical reaction occurs when there is an encounter in the interphase, Figure 3a, or one reactant is transported across the interphase, Figure 3b. The mechanism is much the same when dealing with a water-in-oil microemulsion, the only difference being that here the polar reactant is dissolved in the dispersed phase while the non-polar reactant is in the continuous phase. Because the interphase volume is so large, up to 40% of the total volume, one can expect rapid reaction due to the high probability of reagent encounter. Further modification of reaction rates and pathways can be achieved by 1) varying the amphiphile in such a way as to change the charge gradient across the interphase, 2) adjustment of steric bulk of the interphase through varying of the surfactant concentration or molecular complexity, or 3) through the introduction of a phase transfer catalyst. These are exemplified below.

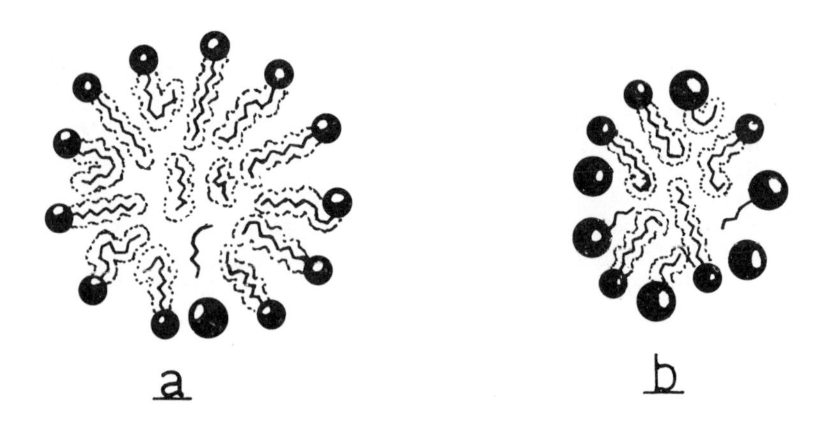

Figure 2. Micellar catalysis: a, reaction of a water-soluble ion with a nonpolar organic compound; and b, reaction of a water-soluble ion with a polar organic compound. Key: ●, *ion;* ∧∧∧∧, *nonpolar organic reactant; and* ●∧∧∧∧, *polar organic reactant.*

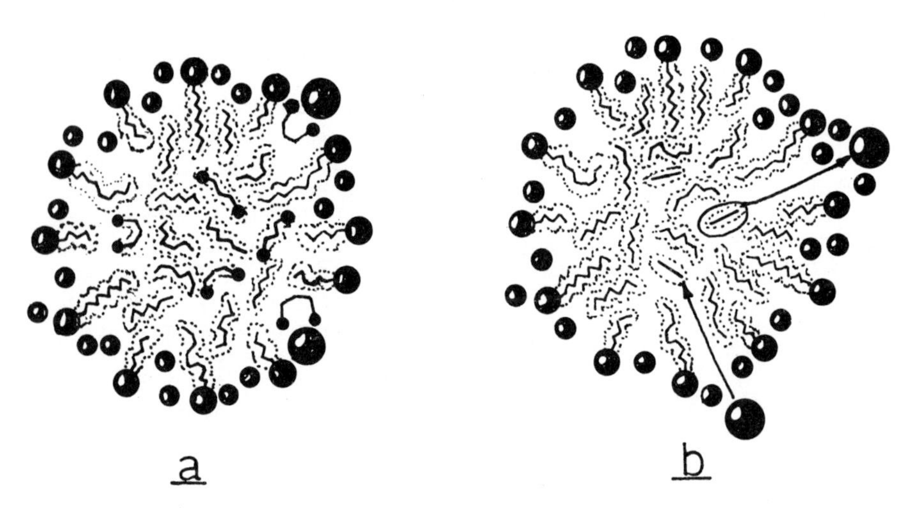

Figure 3. Microemulsion catalysis: a, reaction at an interphase; b, reaction after transport across the interphase. Key: ● , *water soluble ion;* ●, *water;* ∧∧∧∧, *oil phase molecules;* ⌒, *polar organic reactant;* ●∧∧, *surfactant;* ●∧∧∧, *cosurfactant; and* ——, *nonpolar organic reactant.*

Interfacial Effects

The reaction of Cu(II) with meso-tetraphenylporphine, $TPPH_2$,

$$Cu^{2+}_{(aq)} + TPPH_{2(oil)} \rightleftharpoons CuTPP_{(oil)} + 2H^+_{(aq)}$$

is an ideal system with which to probe the nature of microemulsions and to ascertain the utility of microemulsions in modifying reaction rate and pathway. The $Cu(H_2O)_x^{2+}$ ion is soluble only in the aqueous phase while $TPPH_2$ is insoluble in water. As a consequence, reaction must occur in the interphase or some mechanism must be invoked which permits movement of a reactant from one phase to the other.

Studies conducted in our laboratories on the metalation reaction(5,6) have involved two types of microemulsions: detergentless microemulsions composed of water, toluene and 2-propanol and microemulsions of the same formulation but with small amounts (10^{-3}M) of added surfactant. Using these systems both the role of surfactant gegenion and the effect of medium composition on rate and mechanism have been investigated.

The results are tabulated in Table I.

Table I

Observed Pseudo First - Order Rate Constants in "Surfactant" Containing and "Surfactant"-Free Microemulsions

Type	Surfactant	k_{obs} (hr^{-1})
	None	0.00696+.00070
anionic	Sodium Hexadecylsulfate	0.00413+.00035
cationic	Hexadecyltrimethylammonium Perchlorate	0.00612+.00001
	Hexadecyltrimethylammonium Chloride	0.213 $\pm$.0066
	Hexadecyltrimethylammonium Bromide	0.744 $\pm$.0603

Pseudo first-order constants in Table I were obtained in a microemulsion composed of $0.411X_{TOL}$, $0.186X_{H_2O}$, $0.403X_{IPA}$. The rate law for the reaction in the absence of detergent is:

$$rate = \frac{k[Cu^{2+}]\,[TPPH_2]}{[H^3O^+]}$$

In the presence of hexadecyltrimethylammonium bromide, HTAB, this rate law can be written:

$$\text{rate} = k\ [Cu^{2+}]\ [TPPH_2]\ [HTAB]$$

The pseudo first-order rate constants, Table I, for the detergentless system, with added sodium hexadecylsulfate, SHS, and hexadecyltrimethylammonium perchlorate, HTAP, are much the same, i.e. 0.004-0.007 hr^{-1}. Addition of hexadecyltrimethylammonium chloride, HTAC, or HTAB drastically effects the rate however: k_{obs}(HTAC) ≈30 k_{obs}(HTAP) and k_{obs}(HTAB)≈100 k_{obs}(HTAP). This can be rationalized based on the mechanism diagramed in Figure 4a and 4b. In a detergentless system containing only aqueous copper perchlorate and $TPPH_2$ reaction much occur in the interphase since neither reagent shows appreciable solubility in the other reactants host system. The addition of HTAP does little to alter these conditions. While there undoubtably exists a Stern-like layer, it appears that the concentration of HTAP is insufficient in any given microemulsion droplet for this to be a factor. Addition of SHS causes a slight retardation in k_{obs}. This can be rationalized by noting that the sulfate head groups have some affinity for Cu(II) and may be decreasing the copper mobility through complexation. In any case the affect is not large. When HTAC or HTAB are added it is clear from the rate increases that a mechanism which relys only on a random encounter in the interphase is no longer applicable. Insight into the rate enhancement process can be obtained if we compare the stability for the formation of $CuBr_4^{2-}$ and $CuCl_4^{2-}$. Log β_4 for the reaction:

$$Cu^{2+} + 4Br^- \rightleftarrows CuBr_4^{2-}$$

is 8.92 and, for the analogous reaction involving the chloride ion it is 5.62.

These data are consistent with a mechanism whereby the formation of CuX_n^{2-n} species facilitate the metalation reaction. This could be effected in two ways. First, CuX_4^{2-} is formed, attracted to the cation head groups (but not "bound"), Figure 4b. This would then increase the concentration of Cu(II) in the interphase enhancing the probability of an encounter with a $TPPH_2$ molecule. An alternate pathway requires that CuX_2 be the dominant species. This molecule is less polar than $Cu(H_2O)_n^{2+}$ or CuX_4^{2-} and as a consequence can more readily penetrate the toluene continuum. This latter mechanism is phase transfer in nature. Addition of first NaBr then NaBr + HTAB to the detergentless system suggests that both pathways are important. When the Br^- is 5.8x10^{-4}M k_{obs} is found to be 0.0951±0.0126 hr^{-1}, considerably higher than for the reaction in the detergentless system <u>sans</u> NaBr. If 3.5x10^{-4}M HTAB and 2.2x10^{-4}M NaBr are used (total concentration $[Br^-]$ = 5.7x10^{-4}) k_{obs} is 0.172 ± 0.025, a factor of 2 greater than that observed with NaBr alone. Since there is no surfactant head group when only NaBr is used it is likely that transport is

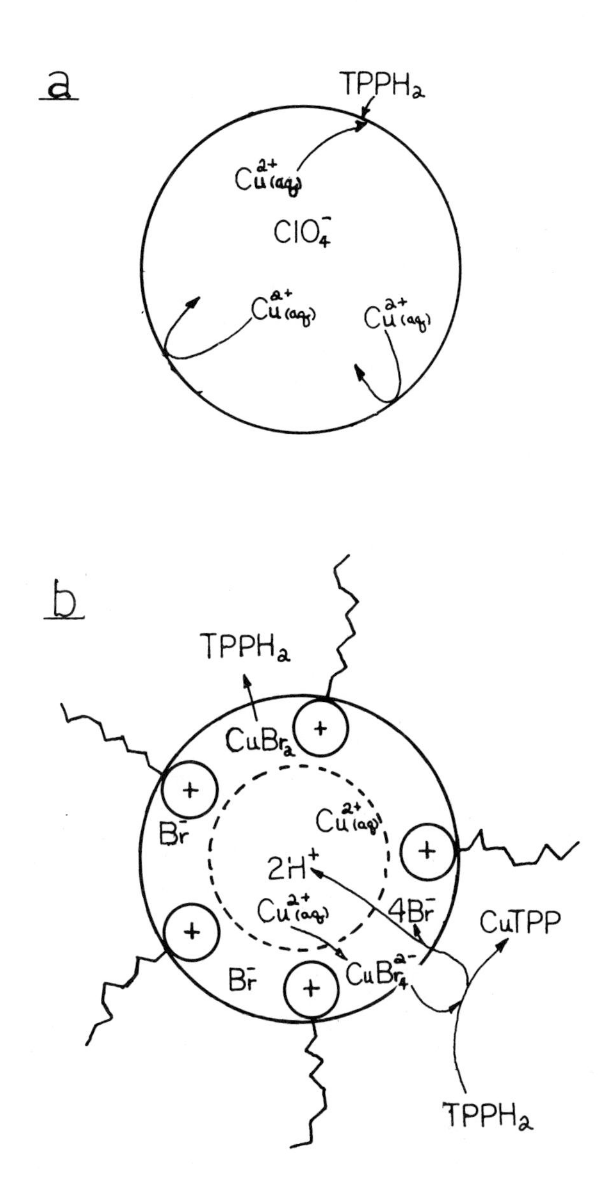

Figure 4. Reaction of Cu^{2+} with $TPPH_2$: a, in the absence of added halide; and b, in the presence of added halide.

effected by $CuBr_2$. On the other hand addition of HTAB does increase the rate so it is probable that the $CuBr_4^{2-}$ species are also important.

An investigation of the rate of metalation as a function of solution composition in a detergentless system is also very instructive. The rate of reaction varies little, Figure 5, along the X_{H_2O}=0.2 isopleth while the solution composition lies in the microemulsion region. Once into the "small aggregate" region, however, k_{obs} increases dramatically and continues to increase into the ternary solution formulation region. The principal (and only discontinuous) change, which occurs in leaving the microemulsion region, is coincident with a breakdown of the interphase. This result is a vivid demonstration of the effect of the interphase in the control of the rate of reaction.

The effect of the presence of the microemulsion interphase has also been demonstrated in a study of the base hydrolysis of long chain esters in a water disperse hexane, water, 2-propanol microemulsion(7). Because of the hydrophobic nature of such esters as stearate, laurate, and caprylate, attempts have been made to enhance their rate of hydrolysis in aqueous solution both through the addition of a phase transfer catalyst to a two phase system and by the introduction of micelles. The maximum rate obtained in micellar solution when the reactant was the laurate ester, was 0.26 min^{-1} (8) while under normal conditions for phase transfer catalysis the yield of an ester hydrolysis reaction is ∿ 35%(9). In contrast if the same reaction is carried out in a hexane, water, 2-propanol microemulsion the yield is >98%, with a rate as high as 0.4 min^{-1}. When studies are carried out along an isopleth of constant mole fraction water the rate of hydrolysis changes in a regular manner throughout the microemulsion region, Figure 6, but a discontinuity occurs at solution compositions which correspond to the pseudo-phase boundry. Though not as dramatic an effect as was observed in the studies on the metalation of meso-tetraphenylporphine this behavior again demonstrates the importance of the microemulsion interphase on chemical reactivity.

Interestingly enough, interfacial environment has little effect on the formation of transition metal complexes of N^{α}-dodecanoylamino alcohols(10-12). In a series of studies on Cu(II) complexes with the surface active N^{α}-dodecanoyl histidinol,-lysinol,-glutaminol,-methioninol and -tryptophanol in hexane, water, 2-propanol microemulsions it was found that the formation constants varied little from those obtained in aqueous solution. Further, where it was possible to elucidate structure the coordination geometry was the expected one based on analogy with similar non surface active ligands in aqueous media.

Microemulsions in Chemical Synthesis

Detergentless microemulsions would appear to have considerable

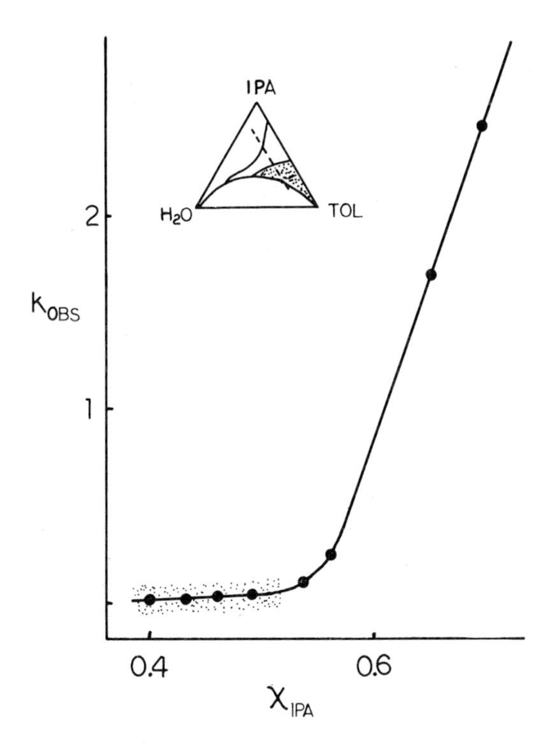

Figure 5. Rate of metalation as a function of a reaction medium composition: microemulsion region is stippled.

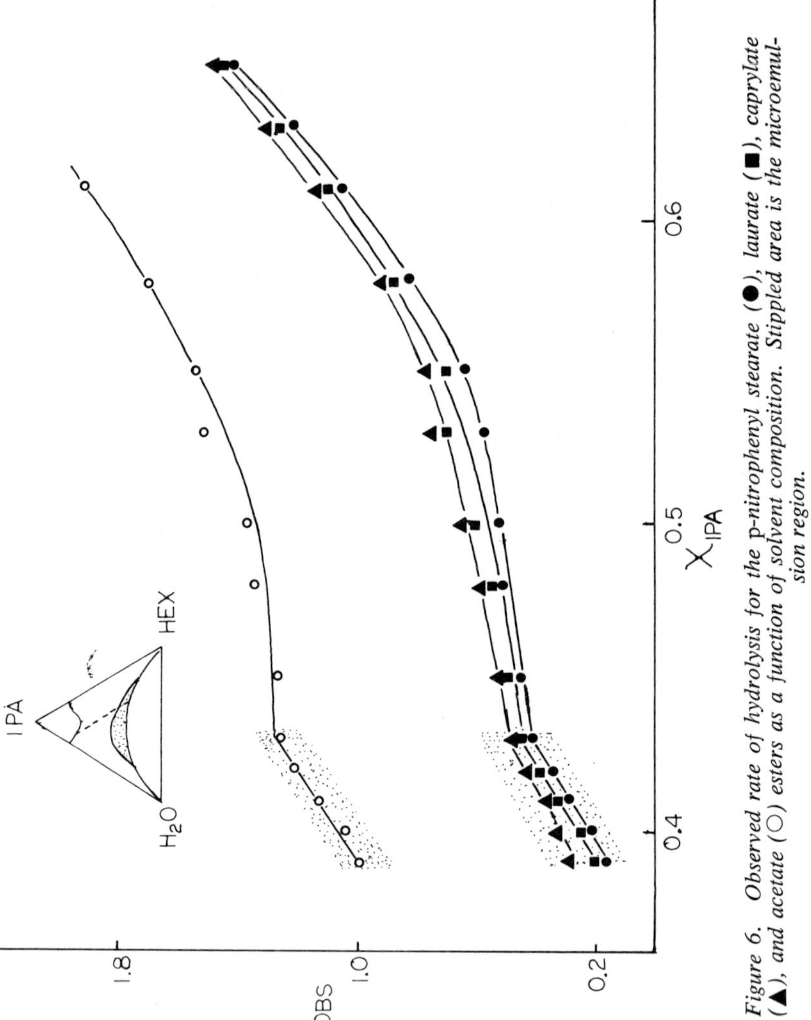

Figure 6. Observed rate of hydrolysis for the p-nitrophenyl *stearate* (●), *laurate* (■), *caprylate* (▲), *and acetate* (○) *esters as a function of solvent composition. Stippled area is the microemulsion region.*

potential for utilization as media for chemical synthesis. As noted earlier not only does the presence of a very large interfacial area enhance the probability of reagent encounter but purification is simplified when compared to a micellar system or when phase transfer catalysts are employed.

The efficiency of detergentless microemulsions in promoting the formation of macrocylic lactones has been studied in our laboratories(13). The two most direct routes to the formation of large macrocyclic lactones are the acid "catalyzed" esterification of ω-hydroxyalkanoic acids and the cyclization of potassium salts of ω-bromoalkanoic acids. In both reactions a competitive pathway yields polymeric material and as a consequence high dilutions are employed with the attendant extended reaction times.

Utilization of microemulsions would appear to be one method by which the polymerization problem might be reduced or even eliminated. In a water-in-oil microemulsion we would expect the ω-hydroxy- and ω-bromoacids to be compartmentalized on a molecular basis i.e. an average of one molecule per drop up to some concentration, then two per drop, etc., and movement between drops inhibited. As a consequence since the base is soluble in water and the acid likely located in the interphase we would expect that the chance of ring closure before dimerization, trimerization, etc., would be greatly enhanced over that existing in homogeneous media.

Using a toluene based detergentless microemulsion to investigate the cyclization of 12-hydroxyoctadecanoic, 15-hydroxypentadecanoic and 16-hydroxyhexadecanoic acids it was found possible to increase the concentration 40-fold and reduce the reaction time to 14 hours (as opposed to days) while obtaining 20% yield of lactone. The biggest deterrent to higher yields in a detergentless system appears to be formation of the 2-propyl ester which appeared as 40% of the final product. (If the analogous reaction is run in a mixture of water and 2-propanol the result is 50/50 ester/polymer but no lactone).

Utilization of 11-bromoundecanoic and 15-bromopentadecanoic acid eliminated the problem of ester formation. After reacting a $5x10^{-3}$M solution of the 11-bromo acid with KOH for a period of one day, 25% lactone and 18% polymer were isolated. The remaining material was recovered as unreacted bromoacid. While neither of the results is spectacular they do demonstrate the potential utility of microemulsions in helping to minimize the effects of an unwanted competing reaction.

The palladium catalyzed formation of ketones from long chain α-olefins has also been investigated(14):

$$R - \underset{H}{\overset{H}{C}} = CH_2 \underset{}{\overset{CuCl_2,\ PdCl_2,\ O_2}{\rightleftharpoons}} R - \underset{\underset{O}{\|}}{C} - CH_3$$

The reaction conditions used were quite mild, e.g. 1 atm O_2 ambient temperature. Reaction rate was studied both as a function of olefin chain length and as a function of solvent composition. It was found that in a microemulsion the rate of reaction decreased with increasing chain length. Catalyst turnover rates, after two hours, were C_{10}=19.1, C_{14}=15.7 and C_{18}=12.4. In the ternary solution, the order of rates was the same but the absolute rates were some 10% - 30% lower. The specificity for the methyl ketone was higher when the reactions were run in a microemulsion rather than in a ternary solution (96% - 99% methyl as compared to 84-86% methyl). It was also found that the longer the chain length the higher the specificity for the methyl ketone. This latter result is likely due to the difficulty experienced by an internal olefin in penetrating the polar interphase to a depth which will allow inter-action with the Pd(II) catalyst.

Summary

Microemulsions appear to have broad applicability for use in enhancing chemical reactivity and modifying reaction pathway. The mechanisms by which this occurs in reactions thus far studied can be understood in terms of reactant partitioning, interphase encounter and phase transfer catalysis.

Acknowledgement

The authors gratefully acknowledge the support of NSF grant No. CHE 7913082, and CHE 8025726.

Literature Cited

1. Holt, Smith L. J. Dispersion Sci. Tech. 1980, 1, 423.

2. Smith, Garland D.; Donelan, Colleen E.; Barden, Roland E. J. Colloid Interface Sci., 1977, 60, 448.

3. Keiser, B. A.; Varie, D.; Barden, R. E.; Holt, S. L. J. Phys. Chem., 1979, 83, 1267.

4. Lund, Gary; Holt, Smith L. J. Amer. Oil Chem. Soc. 1980, 264.

5. Keiser, Bruce; Holt, Smith L.; Barden, Roland E. J. Colloid Interface Sci., 1980, 73, 290.

6. Keiser, Bruce; Ph.D. thesis, Univ. of Wyoming, 1979.

7. Borys, N. F.; Holt, S. L.; Barden, R. E. J. Colloid Interface Sci., 1979, 71, 526.

8. Friberg, S.; Ahmad, S. I. J. Phys. Chem., 1971, 75, 2001.

9. Starks, C. M. J. Amer. Chem. Soc., 1971, 93, 195.

10. Smith, Garland D.; Garrett, Barry B.; Holt, Smith L.; Barden, Roland E. J. Phys. Chem., 1976, 80, 1708.

11. Smith, Garland D.; Garrett, Barry B.; Holt, Smith L.; Barden, Roland E. Inorg. Chem., 1977, 16, 558.

12. Smith, Garland D.; Barden, Roland E.; Holt, Smith L. J. Coord. Chem., 1978, 8, 157.

13. Gonzalez, Alvaro; Holt, Smith L. J. Org. Chem., 1981, 46, 2594.

14. Murphy, John; Holt, Smith L. to be published.

RECEIVED August 11, 1981.

11

Inorganic Reactions in Microemulsions

R. A. MACKAY[1] and N. S. DIXIT
Drexel University, Department of Chemistry, Philadelphia, PA 19104
R. AGARWAL
Douglas College, Rutgers University, Department of Chemistry, New Brunswick, NJ 08903

The kinetics of the quinoline-promoted incorporation of copper(II) ion by tetraphenylporphine have been examined in a mineral oil in water microemulsion stabilized by the anionic surfactant sodium cetyl sulfate and 1-pentanol as cosurfactant. A first order dependence of the rate on the quinoline concentration is observed, as compared with a second order dependence in a similar benzene in water microemulsion. The nature of the oil also has a significant effect on the electrochemical reduction of Cu(II), the half-wave potential($E_{1/2}$) being about 0.9 volts more negative in the mineral oil microemulsion. The addition of quinoline causes a positive shift in $E_{1/2}$ which is ascribed to the formation of a four coordinate Cu(I) complex. Although aqueous inorganic ions are normally repelled by a microdroplet interface of the same charge, it is found that cadmium(II) ion is bound to a droplet stabilized by the cationic surfactant cetyltrimethyl ammonium bromide. This behavior is interpreted as arising from the formation of anionic species such as $CdBr_4^{2-}$ in the Stern layer.

In recent years there has been increasing interest in the aplication of organized media such as micelles (1), vesicles (2), liquid crystalline phases (3) and microemulsions (4) to the study of chemical reactions. Most of the reaction systems examined have been organic reactions, and very few inorganic reactions (5,6,7) have been investigated in microemulsions. However, the first reported use of oil in water (O/W) microemulsions as reaction media involved the incorporation of copper(II) by tetraphenylporphine (8). More recently, electrochemical reactions involving species such as cadmium(II), thalium(I), ferricyanide and ferrocyanide have been carried out in order to investigate the transport prop-

[1] Author to whom correspondence should be addressed.

0097-6156/82/0177-0179$05.00/0

erties of ions in both ionic and nonionic O/W microemulsions (9). These latter studies were not directly concerned with the chemical behavior of these ions, but rather focused on their diffusion coefficients. Nonetheless, electrochemical measurements are also capable of providing information on the electron transfer process at the microemulsion-electrode interface as well as information on the binding and complexation of ions in the surface region of the microdroplet.

In the earlier study of copper tetraphenylporphine (CuTPP) formation in an anionic O/W microemulsion, the rate of reaction was greatly accelerated by the addition of quinoline. We have therefore extended our electrochemical measurements to include studies of the copper(II) - quinoline system in an anionic microemulsion, supplemented by some additional kinetic data. We report here the results of these studies, as well as some supplementary investigations dealing with the effect of surface charge and the nature of the surfactant counterion in interfacial processes.

Experimental

The kinetics of formation of copper tetraphenylporphine were performed at 23°C by monitoring the disappearance of the 513 nm absorption band of the porphyrin. The detailed procedure has been described elsewhere (8). The exact microemulsion compositions employed, as well as phase maps of the microemulsion systems can be found in the original paper by Letts and Mackay (8). The components are simply mixed in any order to produce the microemulsions. Formation is spontaneous and the input of mechanical energy is not necessary. However, stirring is usually employed to speed mixing. The electrochemical measurements were performed with a Beckman Electroscan 30 using a three electrode system and a three compartment cell. The working electrode is dropping mercury, and the auxillary and reference electrodes are platinum and saturated calomel (SCE), respectively. All potentials reported in this paper are vs, SCE. Drop times and mass flow rates of 4-5 seconds and 1.6 - 1.7 mg s^{-1} were employed. The supporting electrolyte was 0.05 M $LiClO_4$. However, at the microemulsion water contents employed (c.a. 60%) the same results were obtained both with and without the supporting electrolyte due to the high concentration of ionic surfactant.

Results and Discussion

Metalloporphyrin formation. Our earlier study of metal ion incorporation by TPP was carried out in a benzene in water microemulsion (ME) stabilized by cyclohexanol and a few different surfactants (8). The influence of Lewis bases, quinoline in particular, was studied in the ME system containing (anionic) sodium cetyl sulfate (SCS). For reasons to be discussed below, it was more convenient to perform the electrochemical studies in a mineral oil in water microemulsion stabilized by 1-pentanol and SCS (10). We therefore performed some reaction rate measurements to determine if the kinetics in the mineral oil and benzene microemulsions were similar. Two differences became immediately evident. First, the absolute rate of incorporation is a bit slower in the mineral oil ME. Second, the reaction does not go to completion in the mineral oil ME as it does in the benzene ME, but reaches a position of equilibrium. The equilibrium concentration of porphyrin, for a given initial porphyrin and copper concentration, depends upon the quinoline concentration. This may be understood based on the equilibria of equations (i) and (ii).

$$PH_2 + Cu^{2+} = CuP + 2H^+ \quad (i)$$

$$QH^+ = Q + H^+ \quad (ii)$$

Here PH_2, CuP, and Q represent the porphyrin (TPP), copper porphyrin, and quinoline, respectively. The equilibrium constants for equations (i) and (ii) are designated K and K_a, respectively. It should be noted at the outset that all of the concentrations refer to the overall (analytical) values. The copper(II) ion is confined to the aqueous phase and microdroplet Stern layer, while the porphyrin and quinoline are distributed within the droplet. Thus, K will be proportional to the true equilibrium constant, but not equal to it.

Under our experimental conditions, the initial copper (∿1mM) and quinoline (∿1-10mM) concentrations are in excess of the porphyrin concentration (∿80μM). Thus, taking $[QH^+] = 2[CuP]$, $[Cu^{2+}] = [Cu^{2+}]_o$, and $[Q] = [Q]_o$, a value of K/K_a^2 can be computed. For $[Cu^{2+}]_o = 1.06mM$ and four different values of $[Q]_o$ ranging from ∿1-8mM, a value of $K/K_a^2 = 1.9 \pm 0.6$ is obtained. The value of K_a for quinoline in the microdroplet surface is not known. However, if its value of 1.3×10^{-5} at 25°C in water is used (15) then $K = 3 \times 10^{-10}$. In the absence of quinoline in an unbuffered ME, the equilibrium lies far in favor of reactant and effectively no reaction is observed.

The position of equilibrium varied from about 40-80% reaction, depending upon the quinoline concentration. In the presence of 0.1mM NaOH, for $(PH_2)_o$ ∿ 60μM, the reaction was about 80% complete. Addition of too much NaOH led to slow precipitation of the copper. In order to obtain a rate constant for the forward reaction, log $(A - A_\infty)$ was plotted vs. t. Here, A is the absorbance of a tetraphenylporphin band at 513 nm and A_∞ is the absorbance which would result if all of the PH_2 were converted to CuP. Straight lines were generally obtained over the first 15-30% of reaction, and the slope was taken to be the psuedo first order forward rate constant k_{obs}. The dependence of k_{obs} on the concentration of quinoline and copper is shown in Figures 1 and 2, respectively. The experimental rate law is given by equation (iii).

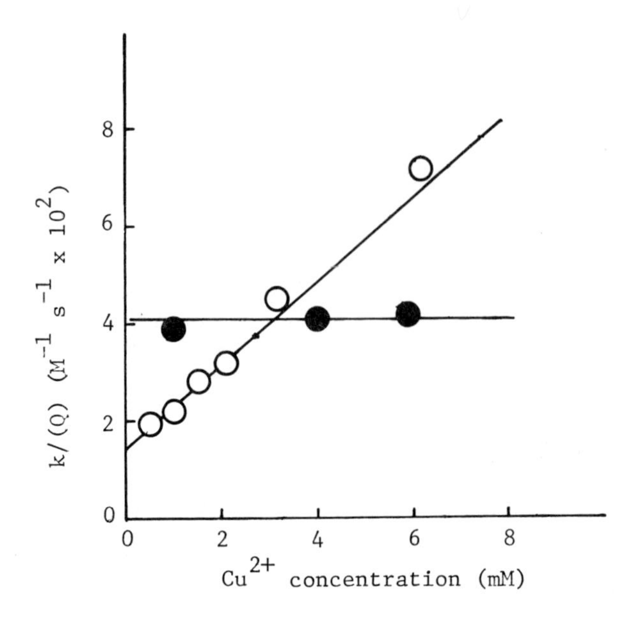

Figure 1. Pseudo first-order rate constant (k) for CuTPP formation divided by the quinoline concentration (Q) vs. Cu(II) concentration in a mineral oil/1-pentanol/SCS (60% water) microemulsion. Key: ○, 0.1 mM NaOH; and ●, no NaOH.

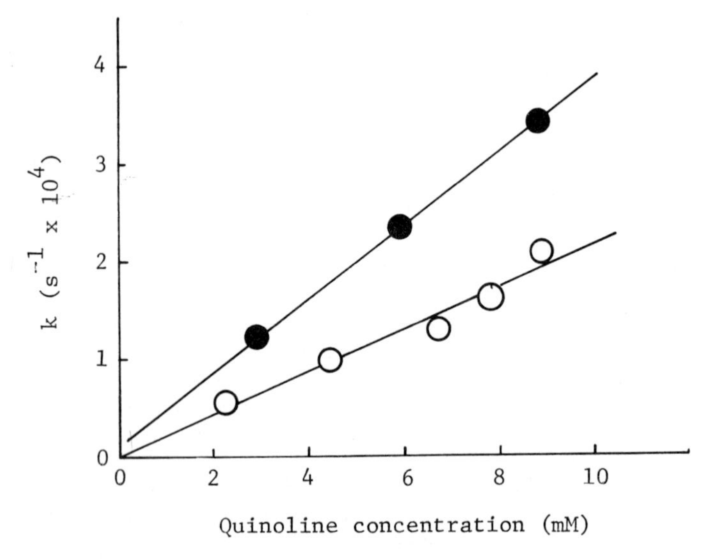

Figure 2. Pseudo first-order rate constant (k) for CuTPP formation in mineral oil/1-pentanol/SCS microemulsion (60% water) vs. quinoline concentration. Key: ○, 0.1 mM NaOH; and ●, no NaOH.

$$-d\ln[PH_2]/dt = k_{obs} \quad \text{(iii)}$$

The dependence of k_{obs} on the copper (M) and quinoline (Q) concentrations is given by equation (iv) and by equation (v) for the microemulsions containing 0.1mM NaOH.

$$k_{obs} = 0.039\ [Q] \quad \text{(iv)}$$

$$k_{obs} = (8.5[M] + .014)[Q] \quad \text{(v)}$$

The concentrations are in mols dm^{-3} and k_{obs} in s^{-1}. These may be compared with the earlier result obtained in a benzene in water microemulsion (8), given by equation (vi).

$$k_{obs} = (1.3 \times 10^4[M] + 42)[Q]^2 \quad \text{(vi)}$$

The second order dependence in the benzene ME was ascribed to the equilibrium formation of a Cu(II) - quinoline complex, which is also in equilibrium with a "sitting-atop complex" (SAT) with the porphyrin. The rate controlling step is then the loss of a proton from the SAT. It is also implicitly assumed that the association constant of the copper-quinoline complex is sufficiently small so that the stoichiometric concentrations of both the metal ion and base may be employed in the rate equation.

It is not necessary to invoke a metal-ligand complex to explain a first order dependence of k_{obs} on the quinoline concentration, since this will arise as a consequence of the removal of a proton from the SAT as the rate determining step. This leads to a first order dependence on both copper and quinoline, which is observed only in the presence of the low level concentration of NaOH. In the absence of NaOH, only a copper ion independent component is obtained which could result from a mechanism involving removal of a proton from the porphyrin, followed by reaction of the porphyrin monoanion with metal ion. The puzzling aspect of this behavior is that one would expect a dependence on added NaOH opposite to that observed.

The solubility of TPP is much greater in benzene than in mineral oil, and it is therefore likely that its average location (10, 11) is nearer to the interface and the copper does not have to be transported (e.g., as a complex) into the droplet interior. Since the microdroplet has a net negative surface charge, it is expected that the local concentration of hydroxide is lower, and that hydroxide cannot effectively penetrate very deeply into the surface region. This is consistent with the effect of hydroxide on an alkylation reaction, to be discussed below. This can account for its failure to increase the rate of the base removal component, but its role in promoting the dependence of k_{obs} on copper ion remains unexplained.

Effect of surface charge. The effect of the microdroplet

surface charge on the rates of a number of chemical processes has been examined. It may cause ejection of an ion of the same charge (12), as well as a lower surface concentration of an aqueous ion of the same charge (4), while the intrinsic rate and equilibrium constants appear to remain relatively unaffected (4, 13, 14). However, little information is available on the relative depth of penetration of simple aqueous ions into an oil microdroplet. That this type of consideration can be important is illustrated by the kinetics of metalloporphyrin formation discussed above.

Some light on this question can be shed by considering the reaction of a substituted pyridine with alkylating agent and hydroxide. The reaction sequence is given below.

NO_2 NO_2 NO_2

CH_2 + RX → CH_2 $\xrightarrow{OH^-}$ CH + X^-

N N^+ X^- N

R R

colorless violet

I II III

The oil-soluble reagent p-nitrobenzylpyridine (I) reacts with alkylating agent RX (here, CH_3I) to produce the pyridinium salt (II), which then reacts with base to form the violet product III. The base must be added after alkylation since it can slowly react directly with I to produce a blue colored product. These reactions were examined in both an anionic (SCS) ME as well as in a cationic CTAB system. The results are given in Table I.

Hydroxide does react directly with I in a cationic ME., but no reaction takes place in the anionic system. Unfortunately, the CH_3I does not alkylate I under these conditions. However, the addition of a small quantity of silver ion results in alkylation, and the hydroxide ion does react directly with the pyrinium

Table I. Reactions of p-nitrobenzylpyridine in O/W microemulsions (*vide text*).

Reagent/ME	(I)[a]	OH^-	(II)[a]	CH_3I	Ag^{+b}	Result
Cationic	x	x				blue color
Anionic	x	x				no reaction
	x	x		x		no reaction
		x	x			violet color (III)
	x	x			x	no reaction
	x	x		x	x	violet color (III)

a. *p*-nitrobenzylpyridine (I) and *N*-methyl-*p*-nitrobenzylpyridinium iodide (II).

b. $AgNO_3$.

ion (II) in the anionic microemulsion. While II is also oil soluble, it would be expected to be located, on the average, closer to the aqueous interface. Therefore it is clear that no reaction between hydroxide and the very oil soluble TPP should take place in an anionic microemulsion.

Interfacial Metal Ion Complexation. Complexation of copper (II) ion by oil soluble ligands has been implicated in metalloporphyrin formation in benzene in water ME's, but not apparently in the mineral oil in water system (*vide supra*). Therefore, an electrochemical study of the interaction of quinoline and copper ion was undertaken.

In the benzene/cyclohexanol/SCS microemulsion, polarographic studies were rendered difficult by high background currents and poorly defined half-waves, usually possessing maxima. However, qualitatively the reduction of copper occurred at about the same potential as in water (c.a. +0.15 volts *vs*. SCE). Addition of quinoline caused a negative shift in half-wave potential ($E_{1/2}$), as normally observed upon complexation. It was therefore concluded that the copper ion remained relatively unaffected by the (monoanionic) alkylsulfate head group of the surfactant in the microdroplet surface, and that it was (probably weakly) complexed to a small extent by the quinoline. The overall equilibrium constants for the formation of mono (β_1) and dipyridine (β_2) complexes of Cu(II) are on the order of 10^2 and 10^4, respectively (15), and it might be expected that quinoline exhibit similar values.

In the mineral oil/1-pentanol/SCS microemulsion, the background current is relatively low and a well defined half-wave with no maximum was obtained for Cu(II). Much to our surprise however, an $E_{1/2}$ of -0.73V *vs*. SCE was obtained. This is a negative shift of almost 0.9V with respect to the value in water or the benzene ME which employed the same surfactant (SCS). Although a single half-wave is apparently observed at [Q] < 10mM (Figure 3), a plot of $i/(i_d-i)$ *vs*. E (Figure 4) shows that there are two processes which are occurring. Here i, i_d, and E are the current, limiting diffusion current, and potential respectively. Since the value of i_d is that expected for the overall two electron reduction of Cu(II) (*vide infra*), the two straight line segments in Figure 4 are treated as two electrochemically irreversible one electron processes corresponding to Cu(II) → Cu(I) at higher (more positive) potential and Cu(I) → Cu(0) at lower potential. From the slopes of the lines in Figure 4, it is possible to calculate the transfer coefficients for the first (α_1) and second (α_2) reduction step, respectively. From the values in Table II, it may be seen that α_1 and α_2 have constant values of 0.39 ± .02 and 0.61 ± .05, respectively, independent of the quinoline concentration. However, the width of the half-wave, as measured by the difference between the potential at 25% ($E_{1/4}$) and 75% ($E_{3/4}$) of i_d, increases with increasing quinoline. Although

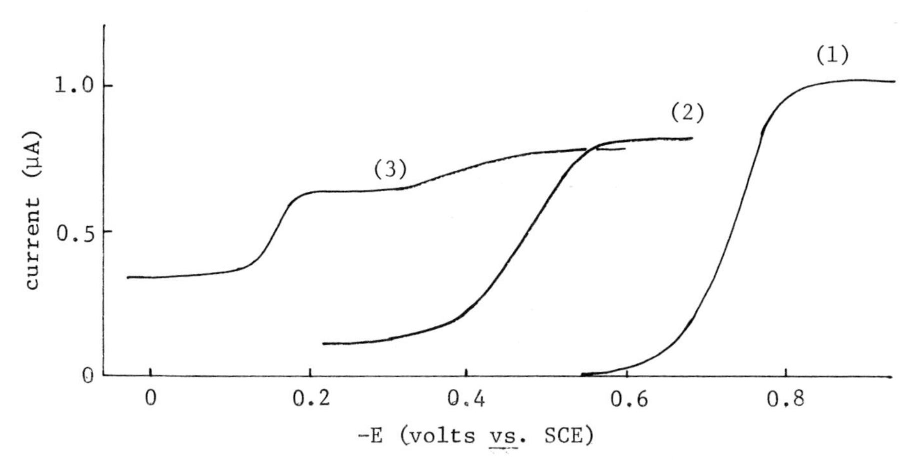

Figure 3. Polarographic half-waves for the reduction of 0.3 mM Cu(II) in a mineral oil/1-pentanol/SCS microemulsion (60% water) containing various concentrations of quinoline (Q). Curve 1, (Q) = 0; Curve 2, (Q) = 32 mM; and Curve 3, (Q) = 147 mM.

Table II. Electrochemical Data for the reduction of Cu(II) as a function of added quinoline in the mineral oil/n-pentanol/SCS microemulsion

[Q]/[Cu(II)][a]	α_1[b]	α_2[b]	$-E_{1/2}$[c]	Δ[d]
0.00	.44	.69	0.733	4
0.27	.37	.59	0.695	2
0.54	.40	.62	0.670	5
0.81	.40	.59	0.658	4
1.08	.37	.64	0.643	5
1.63	.40	.65	0.627	26
2.17	.40	.59	0.612	11
2.71	.39	.60	0.602	14
3.25	.39	.57	0.587	9
4.06	.39	.55	0.570	14
4.86	.35	.62	0.565	22
6.50	.39	.69	0.540	15
8.36	.36	.55	0.525	14

a. The copper and quinoline concentrations varied from 1.4 - 1.8mM and 0 - 11mM, respectively.
b. transfer coefficient (vide text).
c. Half-wave potential (volts vs. SCE).
d. $\Delta = (E_{1/4} - E_{1/2}) - (E_{1/2} - E_{3/4})$, mV.

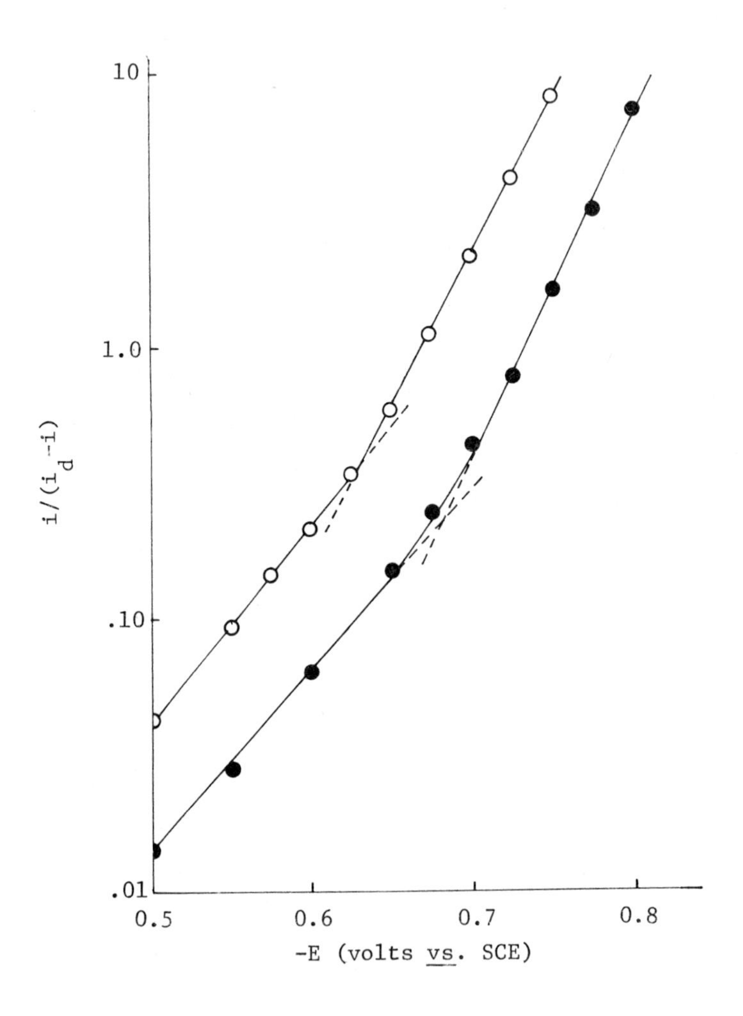

Figure 4. Log $i/(i_d - i)$ vs. potential (E) for the reduction of 1.76 mM Cu(II) in a mineral oil/1-pentanol/SCS microemulsion (60% water). Key: ○, 1.43 mM added quinoline (Q); and ●, Q = 0.

$E_{1/4}$ of i_d, increases with increasing quinoline. Although $E_{1/4}$ is shifting in a positive direction, the values of Δ in Table II indicate that $E_{3/4}$ is shifting negative, but more slowly than $E_{1/4}$ is shifting positive. This suggests that the quinoline is complexing Cu(I) more strongly than Cu(II).

If the quinoline concentration is increased, there appears to be a cationic wave appearing at <0 volts. As the concentration of ligand is increased to ∿150mM, a new (reversible) wave is seen at about -.16V (Figure 3). There is a continuous transition in these features from 10-150mM quinoline, but there is a maximum on the (-.1V) wave at intermediate concentrations of ligand. We do not have a clear interpretation of these results. It may be that as quinoline is added it releases copper from the surface (e.g. the benzene system) since at the higher concentrations it is an appreciable fraction (i.e. - 15%) of the total oil present. The reversible wave which shifts negative with added quinoline may be due to complexation of Cu(II).

If, for the first electron transfer step, Cu(II) → Cu(I), in the half-wave obtained at low (≤10mM) quinoline concentration, the positive shift in potential is ascribed entirely to the formation of a Cu(I) - quinoline complex, then equations (vii) and (viii) are obtained.

$$\frac{dE_{1/2}}{d\log(Q)} = \frac{.059}{n}k \qquad \text{(vii)}$$

$$\left[\frac{\partial \ln \frac{i}{i_d - i}}{\partial \ln(Q)}\right]_E = \alpha k \qquad \text{(viii)}$$

The number of electrons transfered is n (here, n = 1), and k is the number of ligands in the complex (a single complex is assumed). The plots corresponding to equations (vii) and (viii) are shown in Figures 5 and 6, respectively. It should be noted that the half-wave potential employed for the first reduction is the $E_{1/4}$ of the complete two electron wave, and i_d is one half the limiting current of the wave. From equation (vii), a value of k = 3.7 is obtained. From equation (viii), using $\alpha = 0.39$ (α_1 from Table II) a value of k = 3.7 is also obtained. This would tend to indicate that a tetracoordinate Cu(I) complex is formed. The stability constant must be relatively low since the straight line segments in Figures 5 and 6 begin at a Q/Cu ratio of 1.6. It may also be noted that quinoline has the same effect on the reduction of Cd(II), except that only a single, irreversible, two electron half-wave is observed (Figure 7). For n = 2, the slopes yield α values of 0.44 and 0.45 for zero and 6.8mM quinoline, respectively.

Finally, an interesting example of complexation has been ob-

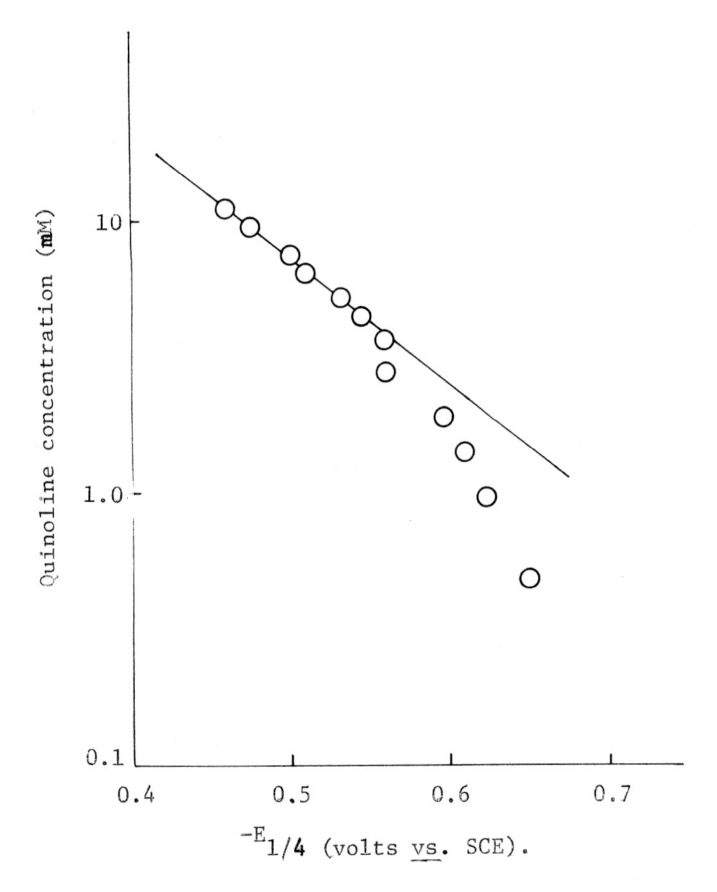

Figure 5. Log quinoline concentration vs. $E_{1/4}$ for reduction of Cu(II) in a mineral oil/1-pentanol/SCS microemulsion (60%) (see *text and Equation vii). Slope of straight line segment at higher quinoline concentration is 4.55.*

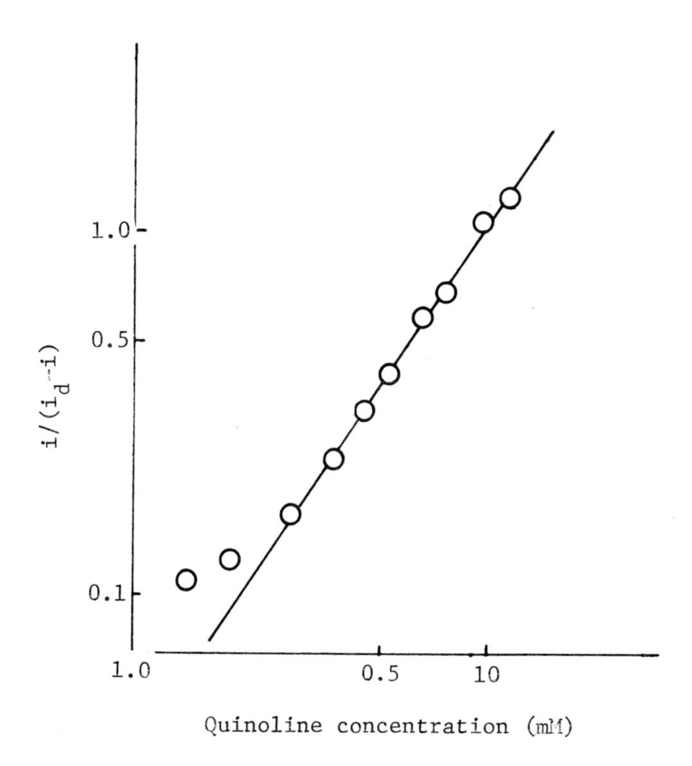

Figure 6. Log $i/(i_d - i)$ at a fixed potential of -0.50 V (vs. SCE) vs. log quinoline concentration (see *text and equation viii). Slope of straight line segment at higher quinoline concentration is 1.46.*

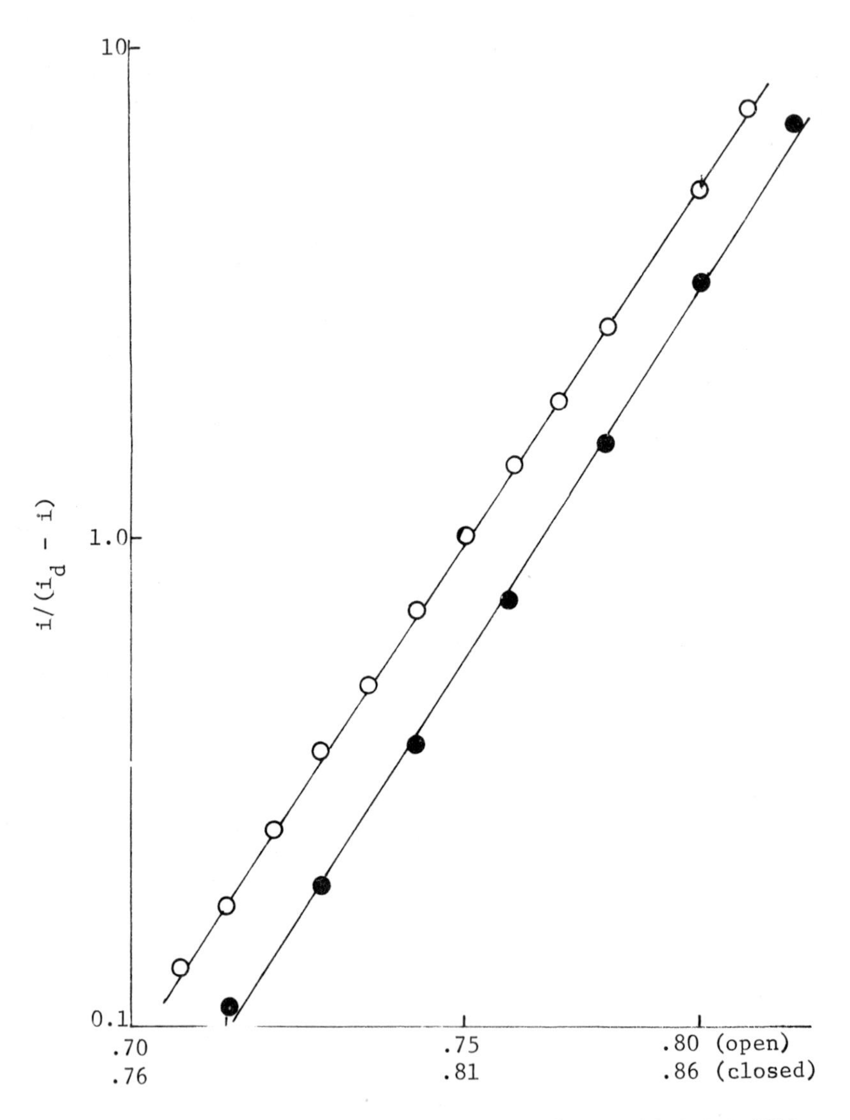

Figure 7. Log $i/(i_d - i)$ vs. potential (E) for the reduction of 2 mM Cu(II) in a mineral oil/1-pentanol/SCS microemulsion (60% water). Key: ○, (upper E scale), 6.8 mM added quinoline (Q); and ●, Q = 0.

served with Cd(II) ion in a hexadecane in water microemulsion stabilized by the cationic surfactant cetyltrimethylammonium bromide (CTAB) and 1-butanol as cosurfactant (16). It might be expected that the Cd^{2+} ion be repelled by the positively charged microdroplet. However, the value of the diffusion coefficient (D) determined from the limiting current of the half-wave for the reduction of Cd(II) is constant upon dilution with water over the range 35-70%. The constant value of $D = 7.2 \pm 0.7 \times 10^{-7}$ cm^2 s^{-1}, indicating that all of the Cd(II) is in fact strongly bound to the microdroplet (9). It may be estimated that the concentration of bromide counterion in the Stern layer of the microdroplet is on the order of 3M. Using the formation constants for complexation of Cd(II) by Br^- in water (15), approximately 90% of the cadmium is present as $CdBr_4^{2-}$, and the remaining 10% as $CdBr_3^-$. Thus, the Cd(II) is bound to the cationic drop as anionic bromide complexes. This conclusion is consistent with the results of a study of copper ion incorporation by TPP in a W/O type microemulsion (6). It was found that the rate of incorporation was affected only by cationic surfactant, and then only when the counterion was capable of complexing the Cu(II) (e.g. - bromide was effective but perchlorate was not). This was ascribed to the formation of CuX_4^{2-} species at the oil-water interface.

Summary

The microemulsion components, particularly the nature of the oil, has been shown to have a dramatic effect on the interaction of metal ions in the microdroplet interfacial region. In particular, the change from benzene to mineral oil causes a change in the quinoline dependence of the rate of metalloporphyrin formation and a 0.9 volt shift in copper(II) half-wave potential. The positive shift in $E_{1/2}$ for Cu(II) upon addition of quinoline is ascribed to the stabilization of Cu(I) via a tetracoordinate complex. It has been shown that simple aqueous ions such as hydroxide can react with species in the Stern layer of a microdroplet of the same charge, but cannot penetrate as deeply into the interface as compared with a droplet having the opposite surface charge. The high concentration of bromide counterion in the Stern layer of a cationic microemulsion droplet results in the binding of Cd(II) in the form of $CdBr_4^{2-}$ ions.

It should also be noted that while very few interfacial inorganic reactions have been examined in microemulsion media, these studies have shown that this is a fertile field for future investigations.

Acknowledgement

The support of the U. S. Army Research Office is gratefully acknowledged.

Literature Cited

1. Fendler, E.J., Fendler, J.H., (1975), "Micellar and Macromolecular Catalysis", Academic Press, New York.
2. Ford, W.E., Otvos, J.W., and Calvin, M., (1978), Nature (London), 274, 507; Fendler, J.H., (1980), Accts. Chem. Rsch., 13, 7.
3. Ahmad, S.I. and Friberg, S., (1972), J. Am. Chem. Soc., 94, 5196.
4. Jones, C.A., Weaner, L.E., and Mackay, R.A., (1980), J. Phys. Chem., 84, 1495.
5. Smith, G.P., Barden, R.E., and Holt, S.L., (1978), J. Coord. Chem., 8, 157.
6. Keiser, B., Holt, S.L., and Barden, R.E., (1980), J. Colloid Int. Sci., 73, 290.
7. Robinson, B.H., Steytler, D.C., and Tack, R.D., (1979), J. Chem. Soc. Faraday Trans. I., 75, 481.
8. Letts, K., and Mackay, R.A., (1975), Inorg. Chem., 14, 2990.
9. Mackay, R.A., (1980), paper presented at meeting on "Microemulsions", Faraday Society, Industrial Division, Cambridge, England.
10. Mackay, R.A., Letts, K., and Jones, C., (1977), "Micellization, Solubilization and Microemulsions", (K.L. Mittal, Ed.), Vol. 2, pp. 801-815. Plenum Press, New York.
11. Jones, C.E., and Mackay, R.A., (1978), J. Phys. Chem., 82, 63.
13. Mackay, R.A., Jacobson, K., and Tourian, J., (1980), J. Colloid Int. Sci., 76, 515.
14. Mackay, R.A., and Hermansky, C., (1981), J. Phys. Chem., 85, 739.
15. "Lange's Handbook of Chemistry", 12th Ed. (1979). (J.A. Dean, Ed.), McGraw Hill, New York.
16. Hermansky, C. and Mackay, R.A., (1979), "Solution Chemistry of Surfactants", (K.L. Mittal, Ed.), Vol. 2, pp. 723-729, Plenum Press, New York.

RECEIVED August 3, 1981.

The Hydrolysis of Chlorophyll a in Detergentless Microemulsion Media

An Initial Study in the Development of a Kinetic Model for the Geologic Transmetalation Reactions of Porphyrins Found in Petroleum

DAVID K. LAVALLEE, EUPHEMIA HUGGINS, and SHELLEY LEE

Hunter College of the City University of New York, Department of Chemistry, New York, NY 10021

The most reasonable mechanism for the conversion of natural macrocylic complexes, the chlorophylls, hemes and hemins, to the nickel and oxovanadium porphyrins formed in petroleum appears to be a sequence of hydrolysis and metallation reactions. The chlorophylls and natural metalloporphyrins are hydrophobic whereas the hydronium ion and typical aquated metal ions are hydrophilic so the hydrolysis and metallation reactions may have taken place in interphase regions. Results for the hydrolysis of chlorophyll a in detergentless microemulsion media consisting of toluene, water and 2-propanol show a rate law that is first order in chlorophyll a and second order in acid concentrations,

$$\text{rate} = k\,[\text{chl a}]\,[H^+]^2$$

with $k = 4.7\times10^3\ m^{-2}s^{-1}$ at χ (toluene) = 0.411, χ (water) = 0.180 and χ (isopropanol) = 0.408, T = 25.0°C and I = 0.10 M over a range of 10^4 in k_{obs}. The rate is strongly dependent on the concentration of toluene giving a relationship of $k_{obs} \propto [\text{toluene}]^{2.5}$ while there is no evident correlation of the rate with water or isopropanol concentrations. The reaction follows the same rate law for a variety of compositions within the microemulsion region and even for compositions for which the structure is no longer a microemulsion.

Introduction

Chlorophylls and iron porphyrins are prevalent in plant and animal matter whereas only nickel (as Ni(II)) and vanadium (as oxovanadium V(IV), V=O) metalloporphyrins are found in petroleum. To determine a plausible reaction sequence for these conversions, we are studying hydrolysis and metallation reactions of metal complexes of pheophytins (the demetallated ligands of chlorophylls) and of porphyrins. The pheophytins and metal pheophytinates, including the chlorophylls and the most abundant natural porphyrins, are highly lipophylic and have very low solubilities in aqueous

0097-6156/82/0177-0195$05.00/0

solutions. Common natural forms of metal ions and the hydronium ion are, of course, hydrophilic with low solubility in nonpolar organic solvents. Since reactions of pheophytins and porphyrins with acid and with metal ions during the geological development of petroleum may have occurred at lipid-aqueous interphase regions; we have chosen to study reactions in nonionic microemulsion media. In this report we discuss the hydrolysis of chlorophyll a in a medium consisting of toluene, water and isopropanol.

Background

While crude oil consists mainly of hydrocarbons which have undergone significant chemical changes that reduce their utility as chemical records of the geologic history of the oil formation process, there are also some stable components of oil that have persisted largely intact for very long periods - the metalloporphyrins. The porphyrin species commonly found in oils are vanadyl and nickel complexes of mainly deoxophylloerythroetioporphyrin, deoxophylloerythin, and etioporphyrin III (or mesoetioporphyrin)- see Figure 1 (2, 3). The similarity of these porphyrins to the common porphyrins found in plants and animals, chlorophyll and protoporphyrin IX (the porphyrin found in hemoglobin, myoglobin and many plant and animal cytochromes) shown in Figure 2, was one of the early indicators of the origin of crude oil. Along the fact that most oil occurs in formation with sedimentary rock, the presence of porphyrins provides very strong evidence for the origin of oil being vegetable and animal matter that has been chemically transformed by the effects of temperature and pressure over long periods of time. Metalloporphyrins are very stable compounds, persisting at temperatures of two hundred degrees Celsius with changes of organic substituents on the periphery of the macrocycle but with retention of the basic macrocyclic structure. They have apparently survived from the original deposition of organic material millions of year ago.

As stated by Constantinides and Arich (4):

> The problem of the origin of the metal-porphyrins is closely related to that of the origin of petroleum and is one of the most basic and interesting questions of petroleum geochemistry. The most probable conclusion seems to be that the nickel and vanadium porphyrin complexes are formed by metal exchange reactions from animal and/or plant metabolic pigments such as hemoglobin and chlorophyll.

These comments have been amplified by Hodgson, Baker and Peake (5):

> Among the multitude of compounds occurring in petroleum there are few which are unique. Of these, the porphyrin com-

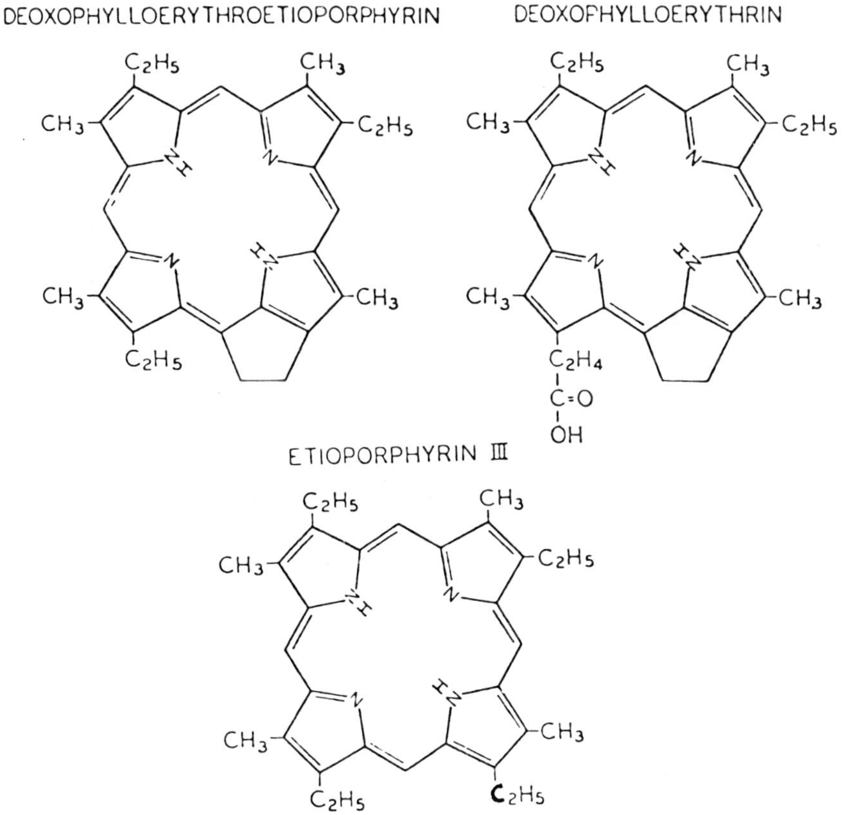
DEOXOPHYLLOERYTHROETIOPORPHYRIN
DEOXOPHYLLOERYTHRIN
ETIOPORPHYRIN III
C2H5
CH3
C2H4
C=O
OH
NH
N

Figure 1. Structures of porphyrins found in petroleum. The species in petroleum are the nickel and oxovanadium complexes.

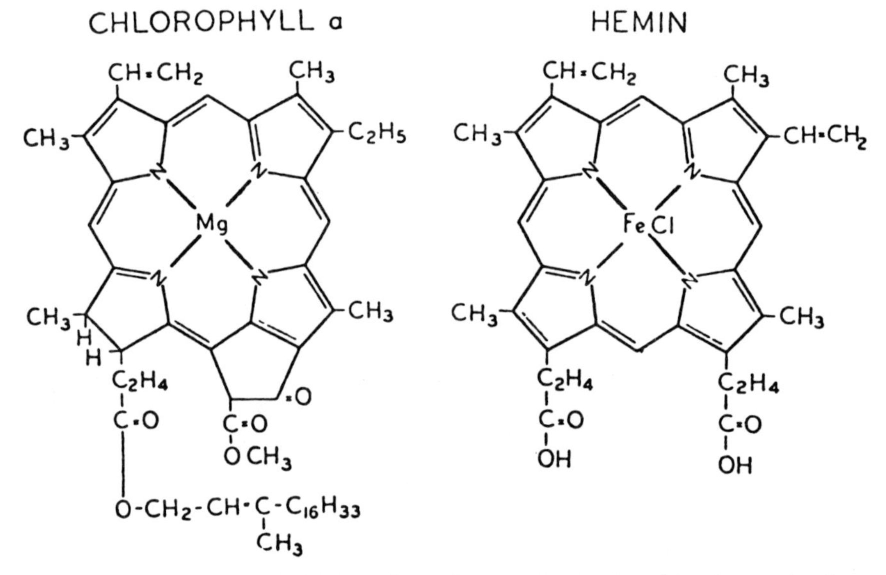

Figure 2. Structures of chlorin and porphyrin molecules found in plant and animal matter. Several different chlorophylls have the same basic ring structure but different peripheral substituents.

pounds are distinctive, offering some hope of defining more clearly the genesis of petroleum. Although there are many possible structures for porphyrins, only a limited number occur in petroleum. However, these few may have had both an active and a passive role to play in the development of crude oil from its source biogenic material. The passive role played by the pigments was that in which the precursor molecules underwent a series of systematic changes until they became the stable and easily recognizable trace compounds of crude oil. In an active sense, the developing pigments may have had a surfactant role to play during the mobilization, migration and accumulation of crude oil hydrocarbons.

The transformation of porphyrin precursors to porphyrins, as well as the occurrence of these compounds in possible petroleum source materials and in petroleum, have considerable geochemical significance in the history of the origin and accumulation of petroleum.

The range of concentrations of porphyrins in oil is quite large - from about 1 ppm in some light oils especially in Texas and Oklahoma to nearly 2000 ppm in heavier crudes from Venezuela (and as high as 0.4 % by weight in some oil shales) (6), with the common concentration of 100-400 ppm. Because of the high extinction coefficients ($\sim 10^5$ M^{-1} cm^{-1}) of both free porphyrins and porphyrin complexes and the very strong fluorescence of free porphyrins (obtained by treating the metalloporphyrins with acid) low concentrations of porphyrins can be quite readily determined. The ratio of vanadium to nickel in the porphyrin fraction can be determined accurately by common techniques such as atomic absorption spectrometry.

Large collections of such data are available (7 - 10). Attempts have been made to correlate the vanadium to nickel ratios (which vary from essentially all vanadium to less than 10% vanadium) with age of the crude oil and limited success - generally with deposits in the same region - has been achieved (5). Further elucidation of the transmetallation mechanism may provide more information for purposes of correlation with history of an oil deposit. It may be found as a result of these investigations that temperature of pH is important in determining the type of metalloporphyrin as well as metal ion availability.

In modelling studies of the chemical transformations of chlorophyll and protoporphyrin IX to the porphyrins found in oils, more attention has been given to the organic chemistry - the changes of substituents at the periphery of the porphyrin ring - than to the way in which these species came to be vanadium and nickel complexes. The presence of the isocyclic ring in deoxophylloerythroetioporphyrin and deoxophylloerythrin points to chlorophyll as precursor while the structure of etioporphyrin (or mesoporphyrin) makes hemin (chloroprotoporphinatoiron(III)) a likely precursor. The conversion of the chlorin ring system found in chlorophyll (one saturated double bond in one of the four pyrrole rings) to the porphyrin system (all unsaturated pyrroles) is readily accomplished by quinones (11). Similarly, bacteriochlorophylls, in which two pyrrolic double bonds have been saturated, can also be converted readily to porphyrins. The other organic reactions that lead to the porphyrins found in oil include saponification of ester substituents, saturation of vinyl groups, reduction of formyl and acetyl groups, reduction of carbonyl groups and decarboxylation of acid groups (6). From the nature of these reactions and the known properties of decaying organic matter, the local environment during the development of oil is presumed to be reducing. Of the metal ions commonly available from ground water and sedimentary rock that can react with porphyrins in contact with water, two have different oxidation states under normal aerobic conditions and under reducing conditions. These are iron and vanadium - two metal ions which are thought to be very important in the conversion of porphyrin and chlorophylls in animal and vegetable matter to the vanadium and nickel porphyrins found in petroleum. Under reducing conditions of the type presumed for petroleum development, iron is expected to exist as Fe(II) and vanadium as V(III) (vanadium(II) is such a powerful reducing agent that it would be expected to reduce organic compounds whereas V(III) near neutral pH is only a mild recuding agent). Other metal ions that are reasonably abundant and could play a role in natural transmetallation reactions are Cu(II), Ni(II), Zn(II), and Mn(II).

Metallation and Transmetallation Reactions

Several different sequences of reactions may be postulated for the conversion of the magnesium complexes of pheophytins (chlorophylls) and iron complexes of protoporphyrin IX and related porphyrins (hemes and hemins) into the nickel and vanadium porphyrins found in petroleum. One possible reason for the isolation of only the nickel and oxovanadium metalloporphyrins is that only they were resistant to degradation. While studies of Hodgson do indicate that complexation of vanadium and nickel do impart added thermal stability to porphyrins (12), Berezin has found that complexation of other metal ions such as cobalt and copper also imparts added thermal stability (13, 14). In addition, Hodgson's study indicates that relatively little thermal degradation of the metalloporphyrins has taken place in most crude oils (which would lead to unbound vanadium and nickel). One would expect that if little degradation of these metalloporphyrins has occurred, complete disappearance of other metalloporphyrins by thermal degradation is an unreasonable assumption.

A second type of mechanism would be the reductive or oxidative demetallation of intermediate metalloporphyrins, allowing nickel and vanadium to complex. From the electrochemistry of porphyrins, and solution studies involving oxidizing and reducing agents, (15) however, it would appear that the only metalloporphyrins which are involved in petroleum maturation that would be susceptible to demetallation under mild conditions by this mechanism are oxovanadium porphyrins (16), which are, in fact, one of the two types of complexes found. It would appear then that none of the intermediate species would be affected by this mechanism.

A third general mechanism is direct transmetallation in which a metal atom displaces a metal atom that is bound to a porphyrin or pheophytin ligand. It could not be necessary for nickel or vanadium to directly displace the magnesium of chlorophylls or the iron of hemes of hemins but a series of transmetallation reactions could occur for which the termination steps involved the nickel and vanadium reactions. The kinetics of transmetallation reactions of several metalloporphyrins have been interpreted in terms of such a direct displacement mechanism (17, 18, 19). The high reactivity of copper (II) as the displacing metal ion in these reactions and relatively low reactivity of nickel (II) implies that at least some copper porphyrin complexes should be observed in petroleum. In addition, the direct (or associative) mechanism for transmetallation has only been postulated for cases involving metalloporphyrins in which the metal ion is significantly out-of-plane (Hg(II), Pb(II), Cd(II) and Zn(II)) and which are also of low stability with respect to acid hydrolysis.

An alternative mechanism for transmetallation is one which involves acid catalyzed dissociation followed by competitive metallation (20, 21, 22). In this process, the metal ion in the metalloporphyrin is displaced by protons. At relatively low acid concentrations (pH $\geq$ 4) expected for ground waters and most environments of developing petroleum deposits, the predominant form of the demetallated porphyrins and pheophytins would be the neutral free base. The metal ions in the vicinity of these free base species could then compete with each other to form complexes. The hydrolysis - competitive complexation sequence would be repeated until the metal ion or ions which are most resistant to acid hydrolysis have bound all the available ligands. The stabilities of metalloporphyrins toward displacement of the metal ion by acid were reported many years ago by Caughey and Corwin (23) and it was found that acid dissociation rates were in the qualitative order Ni(II) < Cu(II) < Co(II) << Zn(II). From a number of qualitative studies Falk and Buchler have constructed a stability series for resistance to acid dissociation as : Ni(II) > Co(II) > Cu(II) > Fe(II) > Zn(II) > Mg(II) with no mention of oxovanadium complexes (24, 25). Although the relatively high stability of Ni(II) porphyrins toward acid displacement might be taken as presumptive evidence in support of the indirect transmetallation sequence involing acid hydrolysis, data of Berezin and Drobysheva must also be considered. Berezin and Drobysheva have studied acid hydrolysis of metal complexes of pheophytin a in mixed solvent systems of ethanol, glacial acetic acid and sulfuric acid. They suggest a stability order of Fe(III) > Cu(II)>

Ni(II) > Co(II) > Zn(II) (13, 14, 26, 27). Considering the concentrations of metal ions which react with porphyrins and pheophytins that are typical of groundwater (in ppm) Fe(0.23), Mn(0.009), Cu(0.007), Ni(0.004) and V(0.001), in sedimentary rock such as graywackes and shale, respectively (Fe(3.84×10^4, 4.8×10^4), Mn (750,850), Cu(45,45), Ni(40,68) and V(67,120)) and in marine sediments (like those of shale, all data from reference 28), and reaction rates typically found in homogeneous solutions (29, 30, 31) one would expect to find Cu(II) porphyrins in petroleum if Berezin's order dominated the transmetallation sequence.

Several interesting possibilities for unraveling the present ambiguities exist. It is quite possible that the relative acid stabilities for a series of metal ions as pheophytin complexes and as porphyrin complexes are indeed very different and the crucial transmetallation hydrolysis reactions occurred after the pheophytins had been converted to porphyrins. Another interesting possibility to explain the absence of Cu(II) porphyrins in petroleum is that the hydrolysis and metallation reactions did not take place in homogenous solution but in interphase regions and that the kinetics of these reactions are significantly different from those in homogeneous solution. Herein we report on the first stage of our investigation of this aspect of the development of a kinetic model for the metalloporphyrin transmetallation sequence for the formation of the complex found in petroleum: the hydrolysis of chlorophyll in a detergentless microemulsion medium.

Results and Discussion

Holt, Barden, and coworkers have recently reported the characterization of detergentless microemulsion media consisting of toluene, water and 2-propanol and hexane, water and 2-propanol (32, 33, 34). In developing pseudophase diagrams for these systems, they have identified regions of stable water-in-oil microemulsions. Since isopropanol which is present in the interphase region is also quite soluble in both water and hydrocarbons, the structure of these microemulsions is much looser, with a greater average thickness of molecules, than is typical of microemulsions formed with detergents. Keiser and Holt have shown that metallation of a free base porphyrin (tetraphenylporphyrin) occurs readily in the detergentless water-in-oil microemulsion media we have used in this study (35). When they added other surfactants to this medium, they found little effect with nonionic and anionic surfactants but large rate enhancements with cationic surfactants, in sharp contrast with results previously found by Letts and Mackay for oil-in-water microemulsion media (36) and by Lowe and Phillips in aqueous micellar media (37). Hambright has also reported metallation rates for porphyrins in aqueous media containing Tween 80 (38). To avoid the profound influence that ionic detergents can have, we have decided to employ a nonionic medium.

As yet, no kinetic data have been reported for the metallation of pheophytinates or for the hydrolysis of metal pheophytinates or metalloporphyrins in well-characterized microemulsion media.

For this study, we prepared toluene, water, 2-propanol microemulsions in which chlorophyll a (extracted from fresh spinach, 39) was dissolved in toluene and perchloric acid and lithium perchlorate (to maintain the ionic strength at 0.10 M) were dissolved in the aqueous phase. Kinetics were monitored by observing changes in the visible absorption spectrum with a conventional or a stopped-flow spectrophotometer, as appropriate, and the temperature was carefully maintained at 25.0°C. All kinetic runs were pseudo-first-order with the chlorophyll a concentration at less than 10^{-4}M. In all cases, the change in absorbance corresponded to a first-order process for several half-lives. A typical plot is shown as Figure 3. As reported previously (39), at a single microemulsion composition of χ (toluene) = 0.411, χ (H_2O) = 0.180 and χ (2-propanol) = 0.403, the dependence of the observed rate on acid concentration was second order over four orders of magnitude of rate (from k_{obs} = 1.0 to 10^{-4}). The data of Berezin, et. al. (13, 14, 26, 27) have indicated non-integral behavior for acid catalyzed hydrolysis of other metal pheophytinates and the nature of the solvent systems used (eg ethanol, glacid acetic acid) have often made determination of the hydrogen ion activity tenuous and the range of acid concentrations studied were generally very restricted. More extensive studies of acid catalyzed dissociation have been performed with metalloporphyrins, including Zn(II), Mn(II), Cd(II), Fe(II) and Fe(III) complexes (29, 40, 45). The rate laws at high acid are typically second order but at low acid, third order rate laws have been observed. For example, for the acid catalyzed hydrolysis of one of closest analogues to the petroleum porphyrins, etioporphinatozinc(II). the rate law observed by Hambright and coworkers(41) is:

$$\text{rate} = (k\ [\text{ZnP}][\text{HCl}]^3)/(\rho + [\text{HCl}])$$

so that the observed rate appears to be second order at high acid concentration but third order at low concentration. For the chlorophyll hydrolysis reaction we find no evidence whatever of third order behavior. Quite possibly this is due to the presence of one saturated pyrrolic ring in the pheophytin ligand. The mechanism that is proposed for the chlorophyll hydrolysis is shown in Figure 4. This mechanism requires the breaking of two bonds between nitrogen atoms and the magensium and subsequent rapid dissociation (39). This result is quite different from that reported by Berezin for the hydrolysis of chlorophyll a in ethanol/glacial acetic acid mixtures, in which he found no simple dependence on the concentration of glacial acetic acid, but "a parabolic dependence on the square of the glacial acetic acid concentration or an approximately fourth power dependence on H^+_{solu}" (26). The mechanism proposed by Berezin is one in which the

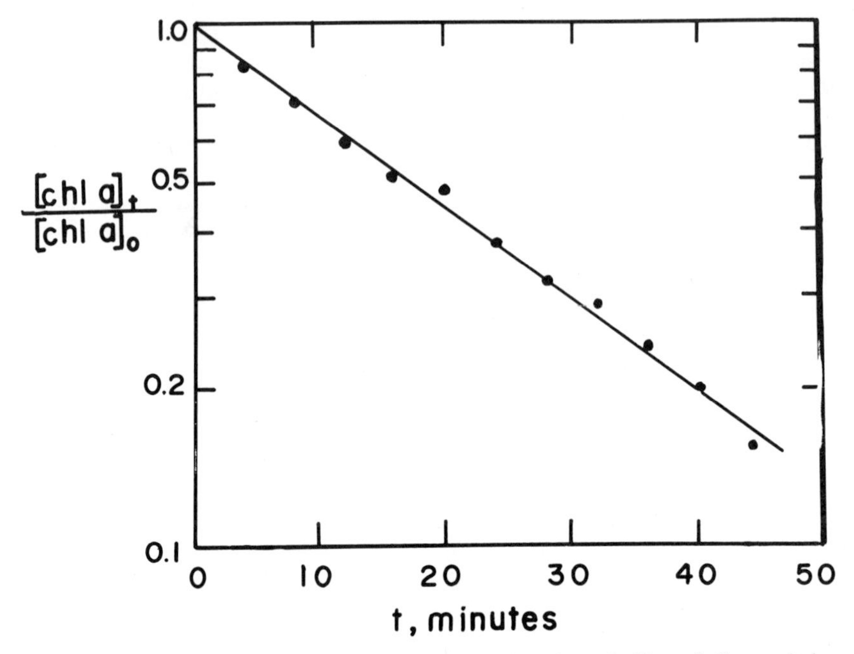

Figure 3. A typical plot of the change in concentration of chlorophyll a as it is hydrolyzed to pheophytin a in microemulsion media.

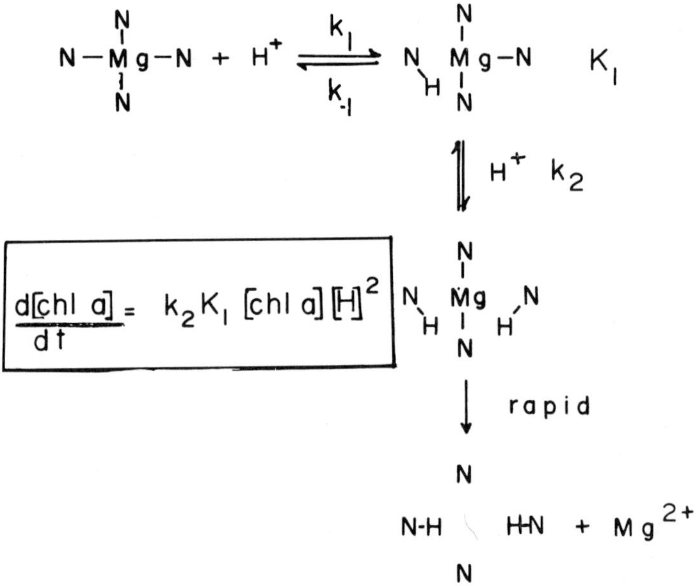

$$\text{N}_4\text{Mg} + \text{H}^+ \underset{k_{-1}}{\overset{k_1}{\rightleftharpoons}} \text{N}_3\text{(NH)Mg} \quad K_1$$

$$\text{N}_3\text{(NH)Mg} \xrightarrow{\text{H}^+,\ k_2} \text{N}_2\text{(NH)}_2\text{Mg} \xrightarrow{\text{rapid}} \text{N}_2\text{(NH)}_2 + \text{Mg}^{2+}$$

$$\frac{d[\text{chl } a]}{dt} = k_2 K_1 [\text{chl } a][\text{H}]^2$$

Figure 4. A mechanistic scheme for the hydrolysis of chlorophyll a that is consistent with the observed rate law.

rupture of the first Mg-N bond is rate determining, a conclusion that is difficult to reconcile with the data. The detergentless microemulsion medium appears to offer a much better system for determining hydrolysis reaction mechanisms.

Table I. Rate Constants For Chlorophyll A Hydrolysis in Media Of Different Composition.

Composition			$k_{calc} = (k_{obs}/[H^+]^2) \times 10^{-2}$, $M^{-2}s^{-1}$			
			$[H^+]$, M			
χ(PrOH)	χ(tol)	$\chi(H_2O)$	0.0078	0.013	0.025	0.031
0.51	0.39	0.094	62	97	100	76
0.64	0.26	0.101	28	23	28	33
0.79	0.17	0.042	7.8	6.9	3.8	4.8
0.85	0.12	0.030	4.7	5.0	5.3	6.0
0.90	0.00	0.10	3.6	2.2	2.7	2.1

It is of interest to us to see how the composition of the microemulsion medium affects the hydrolysis rate. The data in Table I demonstrate that the hydrolysis reaction is second order in acid concentration. It is also quite evident from the table that the hydrolysis rates are indeed sensitive to the composition of the microemulsion. Figure 5 is the pseudophase diagram for the toluene, water, 2-propanol system. We are, of course, dealing with a slightly modified system since the aqueous phase has an ionic strength of 0.10 M. Barden, Holt and coworkers have shown for the rather similar hexane, water and 2-propanol system, however, that addition of NaCl does not drastically alter the pseudophase diagram. Hence, we expect that the compositions used to obtain the data in Table 1 represent all three of the regions of low turbidity, B, C and D. In all cases the media we have used appear transparent. Despite the fact that several different structural regions are represented by the compositions in Table I, there does not appear to be any sharp break in the observed rates with composition but instead a rather gradual change. From the data in Table I, it is evident that there is no simple relationship between the observed rate and the amount of water in the media. The mole fraction of water in all of these compositions is sufficiently small that the propanol concentration and the toluene concentration are highly correlated in an inverse relationship.

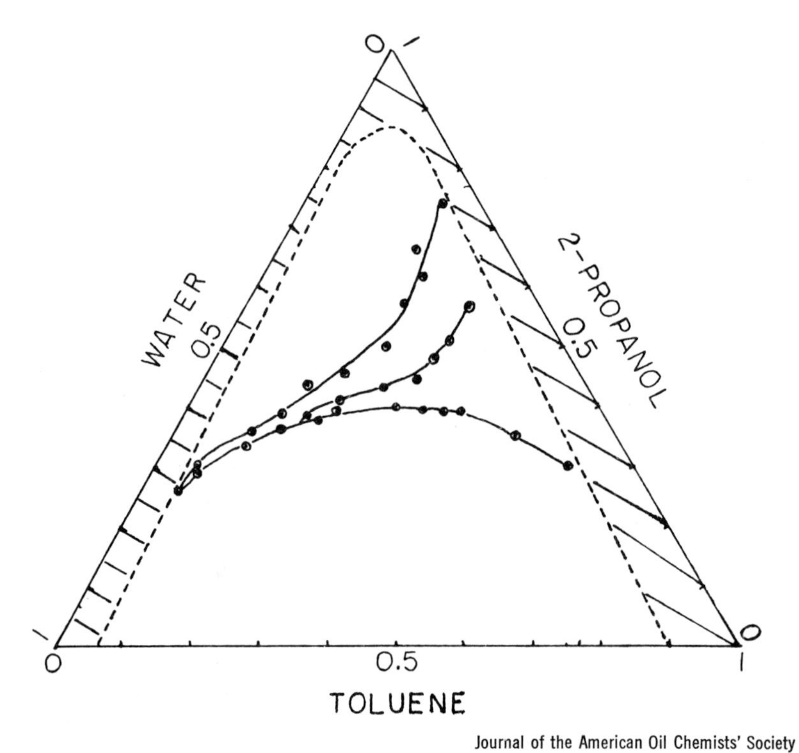

Journal of the American Oil Chemists' Society

Figure 5. The pseudophase diagram for the toluene/water/2-propanol system (34).

Thus, although the rate decreases with increasing propanol concentration and increases with increasing toluene concentration, it is not clear from these data which of the two species is most important in affecting the hydrolysis rate. Further experiments were performed in which the propanol concentration was maintained with water and toluene concentrations varied and with toluene concentration held fixed and water and 2-propanol concentrations varied. When these data are included, the dependence of the rate on toluene concentration (Figure 6) is obvious while the dependence on propanol concentration does not appear to be highly correlated (Figure 7).

In an attempt to determine if a rational relationship exists between the observed hydrolysis rate and the toluene concentration, we attempted a number of correlations to see if surface area of the toluene component or volume available to the chlorophyll might be the crucial factor. The only correlation which we have found is that a plot of the logarithm of the observed rate versus the logarithm of the toluene concentration gives a straight line with a slope of 2.5. Such a result is consistent with an equilibrium involving association of toluene with chlorophyll which increases the reactivity of the chlorophyll toward hydrolysis. This is a linear free energy relationship which would result if, for example, propanol stabilizes the magnesium in the porphyrin by blocking the approach of acid to the nitrogen that bind the magnesium atom. If the toluene displaces the propanol, hydrolysis could be easier. While consistent with a reasonable physical picture, it is certainly a speculative interpretation. We plan to test this hypothesis using other media (for example, the hexane, water, 2-propanol system) which have good solubilization properties for the chlorophyll and acid reactants but for which the hydrocarbon fraction is not expected to associate strongly with the chlorophyll ring system. It should be noted that although a straight line log-log plot is often not a stringent test of a correlation between parameters, the agreement is, in fact, quite remarkable in this case as evidenced from the values calculated from the relationship obtained from the log-log plot and the experimentally determined values of the rate constant (Fig.6). It is interesting that the rate dependence on toluene concentration is the opposite of what might be expected from orientational considerations. It might be presumed that the most favorable orientation for hydrolysis would be one in which the plane of the aromatic ring system of the chlorophyll molecule is parallel to the surface of the water droplet so that the acid could directly attack the coordinating nitrogen atoms without traversing propanol and/or toluene. Increasing the amount of toluene might be expected to reduce the amount of chlorophyll oriented in this manner since the effective concentration of chlorophyll in the interphase region is reduced. Diffusion to the interphase region is likely so fast compared with hydrolysis and the concentration changes of toluene are sufficiently small, however, that no specific steric effect due to the concentration change is evident.

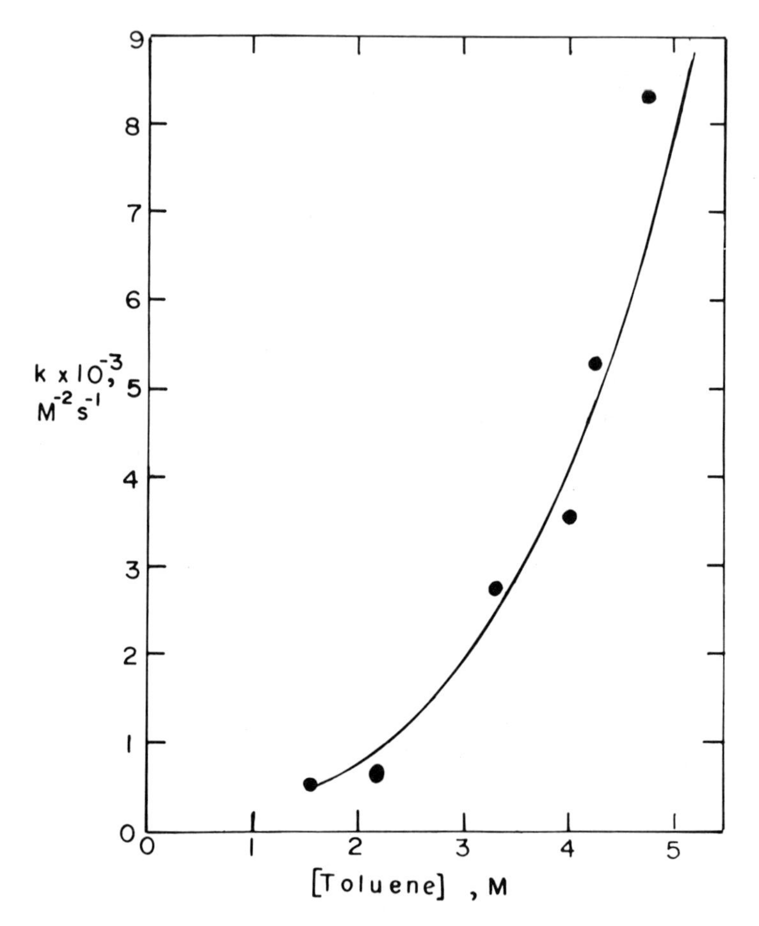

Figure 6. The variation of the third-order rate constant with different concentrations of toluene.

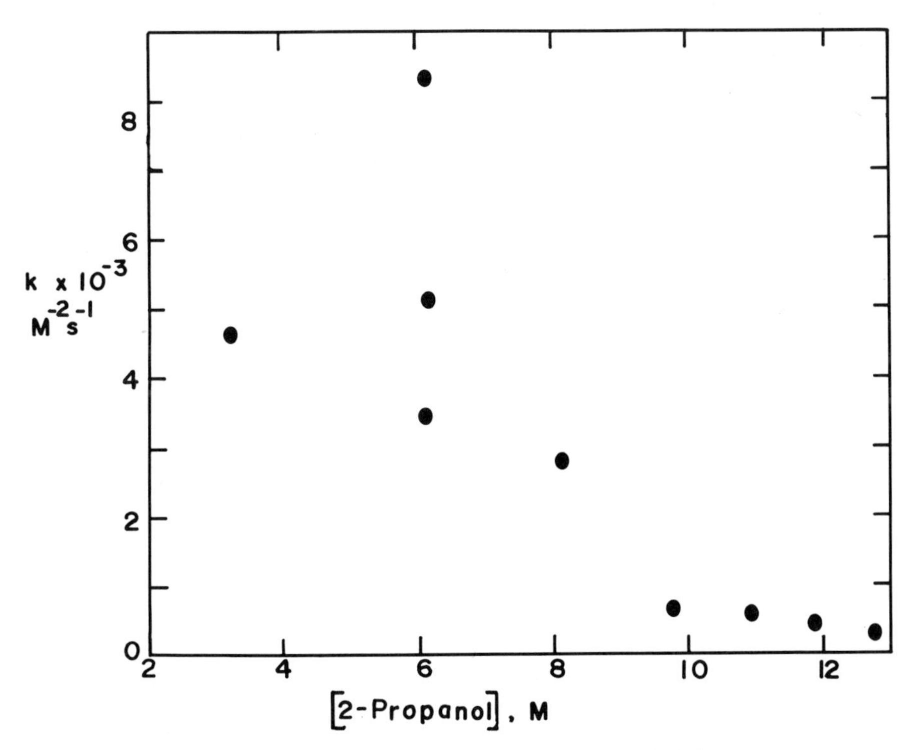

Figure 7. The variation of the third-order rate constant with different concentrations of propanol.

Conclusions

The hydrolysis of chlorophyll a in a detergentless microemulsion medium consisting of toluene, water and 2-propanol is readily monitored giving clean third-order kinetics, first order in chlorophyll and second order in acid concentration, over a wide range of acid concentrations. The reaction is faster in the microemulsion region than in solutions of propanol and water or in regions where the structure consists principally of large hydrogen - bonded aggregates of propanol and water (34). The hydrolysis rate shows a marked dependence on the concentration of the hydrophobic phase, toluene.

Acknowledgements

We are grateful for support of this research by the PSC-BHE grants program of the City University of New York, the NIH-MBS program, the SEED program of the American Chemical Society and Ciba-Geigy Corporation.

Literature Cited

1a. NIH-MBS program participant, b. A.C.S. SEED program participant, cosponsored by Ciba-Geigy corporation.
2. Triebs, A. Angew Chem., 1936, 9, 682-6.
3. Serebrennikova, O.V.; Melkov, V.N. and Titov, V.I., Geohimiya, 1977, 6, 925-31.
4. Shears, B.; Shah, B. and Hambright P. J. Amer. Chem. Soc. 1971, 93, 776-8.
5. Hodgson, G.W.; Baker, B.L. and Peake, E. in "Fundamental Aspects of Petroleum Geochemistry", B. Nagy and U. Colombo, eds., 1967, Elsevier, New York, 177-260.
6. Hodgson, G.W. in "Fundamental Aspects of Petroleum Geochemistry," B. Nagy and U. Colombo, eds., 1967, Elsevier New York, 1-36.
7. Katchenov, S.M. Dokl. Akad. Nauk. USSR, 1977, 232, 456-62.
8. Seber, G.; Weisser, O. and Sesulka, V. Riv. Combust, 1975, 29, 380-6.
9. Dunning, H.N.; Moore, J.W.; Bieber, H. and Williams, R.B. J. Chem. Eng. Data, 1960, 5, 546-9.
10. Arich, G. and Constantinides, G. Riv. Combust, 1960, 14, 695-716.
11. Eisner, U. and Linstead, R.P. J. Chem. Soc., 1955, 3749-54
12. Hodgson, G.W. and Baker, B.L. Bull. Am. Assoc. Pet. Geol., 41, 1957, 2413-26.
13. Berezin, B.D. and Drobysheva, A.N. Russ. J. Phys. Chem., 42, 1968, 1500-4.

14. Berezin, B.D. and Drobysheva, A.N. Russ. J. Phys. Chem., 40, 1966, 386-9.
15. Fuhrhop, J.H. in "Porphyrins and Metalloporphyrins", K.M. Smith, ed., 1975, Elsevier, New York 593-623.
16. Sugihara, J.M.; Branthaver, J.F. and Wilcox, K.W. in "Role of Trace Metals in Petroleum", T.F. Yen, ed., 1975, Ann Arbor Science, Ann Arbor, 183-193.
17. Grant, C. and Hambright, P. J. Am. Chem. Soc., 1969, 91, 4195-8.
18. Stinson, C. and Hambright, P. J. Am. Chem. Soc., 1977, 99, 2357.
19. Baker, H.; Hambright, P.; Wagner, L. and Ross, L. Inorg. Chem., 1973, 12, 2200-2.
20. Lavallee, D.K.; Kopelove, A.B. and Anderson, O.P. J. Amer. Chem. Soc., 1978, 100, 3025-33.
21. Das, R.R. J. Inorg. Nucl. Chem., 1972, 34, 1263-9.
22. Kodama, M. and Kimura, E. J. Chem. Soc., Dalton, 1980, 2447-57.
23. Caughey, W.S. and Corwin, A.H. J. Amer. Chem. Soc., 1955, 77, 1509-13.
24. Falk, J.E., "Porphyrins and Metalloporphyrins," 1964, Elsevier, Amsterdam.
25. Buchler, J.W. in " Porphyrins and Metallporphyrins," K.M. Smith, ed., 1975, Elsevier, New York, 157-232.
26. Berezin, B.D. and Drobysheva, A.N., Russ. J. Phys. Chem. 1970, 44, 1597-1601.
27. Berezin, B.D. and Drobysheva, A.N., Russ. J. Phys. Chem. 1967, 41, 199-202.
28. Wedepohl, K.H. in "Origin and Distribution of the Elements," L.H. Ahrens, ed., 1967, Pergamon Press, New York, 999-1016
29. Hambright, P. in "Porphyrins and Metalloporphyrins" K.M. Smith, ed., 1975, Elsevier, New York,233-278.
30. Longo, F.R.; Brown, E.M.; Rau, W.G. and Adler, A.D., in "The Porphyrins," Vol. 5, D. Dolphin, Ed., Academic Press, New York, 459-71.
31. Schneider, W., Structure and Binding,1975, 23, 123-66.
32. Smith, G.D.; Donelan, C.E. and Barden, R.E., J. Colloid. Interface Sci., 1977, 60, 488-96.
33. Keiser, B.A.; Varie, D.; Barden, R.E. and Holt, S.L., J. Phys. Chem., 1979, 83, 1276-80.
34. Lund, G. and Holt, S.L., J. Am. Oil Chemist's Soc., 1980 57, 264-7.
35. Keiser, B.A.; Holt, S.L. and Barden, R.E., J. Colloid. Interface Sci., 1980, 73, 290-2
36. Letts, K. and Mackay, R.A., Inorg. Chem., 1975, 14,2990-3.
37. Lowe, M.B. and Phillips, J.N., Nature, 1961, 190, 262-3.
38. Hambright, P., Ann. N.Y. Acad. Sci., 1973, 206,443-52.
39. Further details are in: Lavallee, D.K.;Huggins, E. and Lee, S., submitted for publication.

40. Cheung, S.K.; Dixon, L.F.; Fleischer, E.B.; Jeter, D.Y. and Krishnamurty, M., Bioinorganic Chem., 1973, 2, 281-94.
41. Shears, B,; Shah, B. and Hambright, P., J. Am. Chem. Soc. 1971, 93, 776-8.
42. Hambright, P., Inorg. Chem., 1977. 16, 2987-8.
43. Shamin, A. and Hambright, P., Inorg. Chem., 1980, 19, 564-6.
44. Espensen, J.H. and Christensen, R.J., Inorg. Chem., 1977, 16, 2561-4.
45. Reynolds, W.L.; Schufman, J,; Chan. F.and Brasted, R.C.,jun. Int. J. of Chem. Kinetics, 1977, IX, 777-86.

RECEIVED August 4, 1981.

13

Reactions on Solid Potassium Permanganate Surfaces

FREDRIC M. MENGER

Emory University, Department of Chemistry, Atlanta, GA 30322

Oxidations of alcohols on solid potassium or sodium permanganate surfaces take place under mild conditions with high yield and easy workup. Solid sodium permanganate can also oxidize aldehydes and sulfides but not alkenes or alkynes; the solid reagent is thus more selective than the oxidant in solution. Mechanistic aspects (including an observed need for trace quantities of water at the crystal surface, Cu^{+2} catalysis, RCOOH inhibition, crystal deterioration, and reaction intermediates) are still not understood.

Crystallinity is the ultimate in molecular order, so one naturally wonders how reactions at crystal surfaces compare with corresponding "disordered" reactions in solution. Our recent attempts to study this question would, according to the subject of the Symposium, be best introduced with examples of solid state inorganic chemistry. Such examples are, however, rather rare. On the other hand, solid state organic chemistry has been developing rapidly (1, 2); work of others in this area will illustrate the effect of organization in the solid state on reactivity and stereochemistry.

Kornblum and Lurie (3) showed that homogeneous alkylation of sodium phenoxide by allyl bromide in ethylene glycol dimethyl ether (in which both reagents are soluble) gives 99% O-alkylation (Figure 1). In contrast, heterogeneous alkylation of sodium phenoxide suspended in ether produces *ortho*-allylphenol as the major product. Preference for C-alkylation over O-alkylation at the crystal surface most likely arises from the poorly solvated halide ion that would be formed in a solid state O-alkylation (Figure 2). Moreover, covalent bonding to an oxygen at the solid surface would dissipate charge on the oxygen, thereby depriving the sodium ions held in proximity to the oxygen of a counterion. The above difficulties do not apply to C-alkylation at the solid phenoxide surface (Figure 2). The incipient bromide is stabilized by ion-pair

0097-6156/82/0177-0213$05.00/0

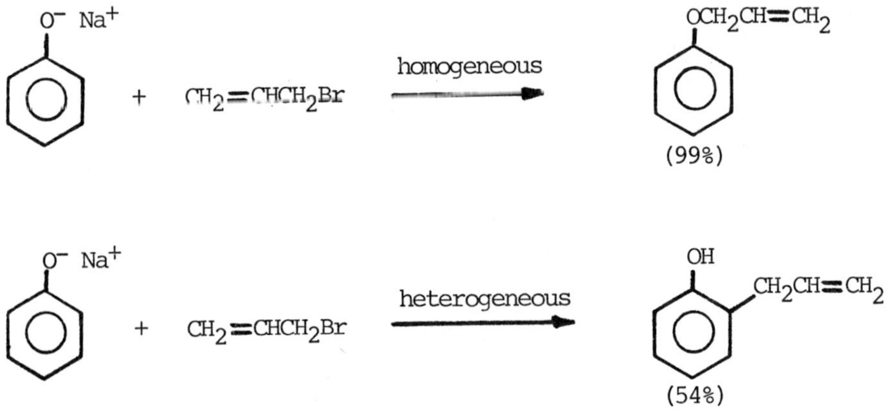

$C_6H_5O^- Na^+$ + $CH_2{=}CHCH_2Br$ —homogeneous→ $C_6H_5OCH_2CH{=}CH_2$ (99%)

$C_6H_5O^- Na^+$ + $CH_2{=}CHCH_2Br$ —heterogeneous→ $o\text{-}HOC_6H_4CH_2CH{=}CH_2$ (54%)

Figure 1.

$Br^{-\delta}$ ⋯ CH_2R ⋯ Na^+ $O^{-\delta}$ Na^+ (phenoxide) → OCH_2R (phenyl ether)

Na^+ $O^{-\delta}$ Na^+ ⋯ $Br^{-\delta}$ ⋯ CH_2R → (cyclohexadienone, CH_2R) —fast→ OH, CH_2R (phenol)

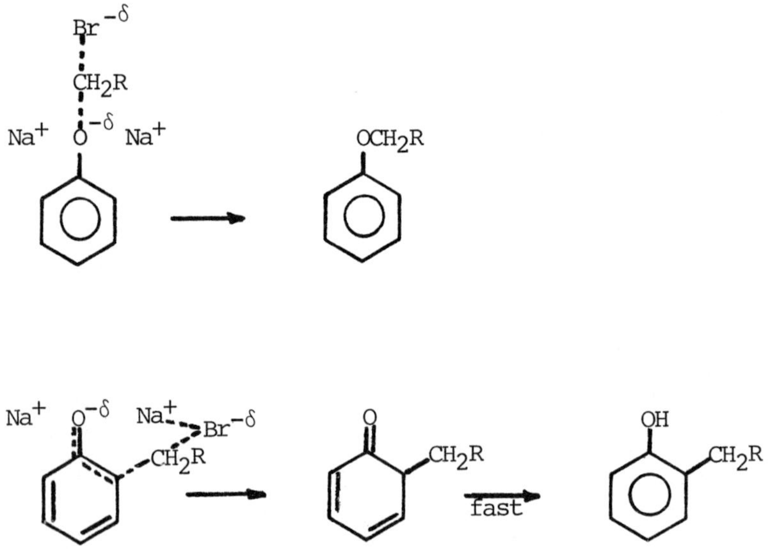

Figure 2.

formation with a sodium ion; charge separation is therefore less unfavorable.

Taylor et. al. (4) found that heating crystalline thallium(I) salts of 1,3-dicarbonyl compounds with alkyl iodides generates C-alkylation products in virtually quantitative yield (Figure 3). None of the common side reactions (dialkylation, O-alkylation, cleavage, etc.) is found. Since an X-ray structure of acetylacetonatothallium(I) shows the carbon backbone located on the crystal surface and the oxygens buried in the interior, the extraordinary reaction specificity probably results from the packing geometry.

Mayer-Sommer (5) alkylated alkali metal pyrazolates both homogeneously and with a suspension of the solid pyrazolate in a solution of the alkylating agent (Figure 4). Two products are possible when R' ≠ R". Heterogeneous conditions, however, lead to the greater product specificity. Thus, when the metal is located nearer to one nitrogen than the other in the crystal, then that particular nitrogen is "blocked" and the other one is selectively alkylated.

Quinkert et. al. (6) photoeliminated carbon monoxide from cis-1,3-diphenyl-2-indanone (Figure 5). In solution, the major product is the more stable trans-substituted benzocyclobutane. However, the cis geometry is maintained in the corresponding solid phase reaction because the crystalline state impedes rotation about carbon-carbon single bonds in the reaction intermediate.

Penzien and Schmidt (7) exposed a single chiral crystal of 4,4'-dimethylchalcone to bromine vapor and showed that the resulting dibromo derivative is optically active (Figure 6). Thus, if a crystal has a chiral structure, then reaction in that crystal can lead to chiral products even though the reactant molecules are achiral. The common textbook statement that reactions of achiral molecules always produce achiral products does not necessarily apply in the crystalline phase.

Curtin et. al. (8) exposed benzoic acid crystals to ammonia gas and generated ammonium benzoate. As shown in Figure 7, the (101) faces of the crystals become opaque while the (001) faces remain clear. The resistance of molecules at the (001) faces to reaction with ammonia is explainable by the relatively unexposed carboxyl groups within this region.

Cheer and Johnson (9) described a system which illustrates the selectivity possible for a reaction on a solid surface. They rearranged an epoxide (Figure 8) homogeneously (BF_3-etherate in CH_2Cl_2) and heterogeneously (on an alumina surface). In the homogeneous reaction, both erythro and threo-epoxide gave predominantly product 1 corresponding to a more favorable aryl migration. The heterogeneous rearrangement catalyzed by alumina behaved quite differently: erythro led to >90% 2 and threo led to >90% 1. This remarkable selectivity was explained by immobile transition states in which both the epoxide and alcohol oxygen atoms of the substrate are fixed to the alumina surface (Figure 9). Such conformational constraints are, of course, not imposed on the molecules in solution.

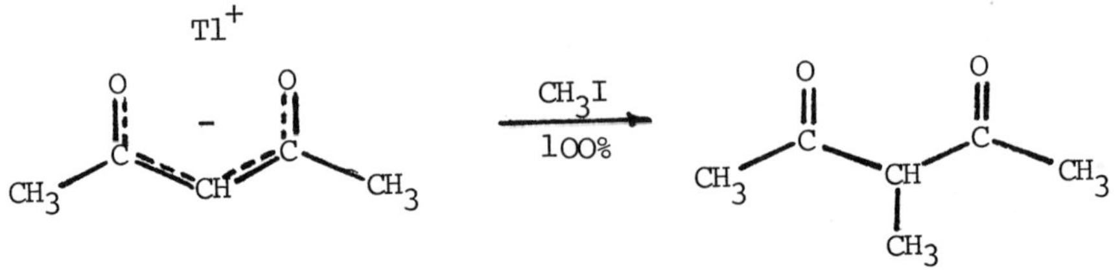

Figure 3.

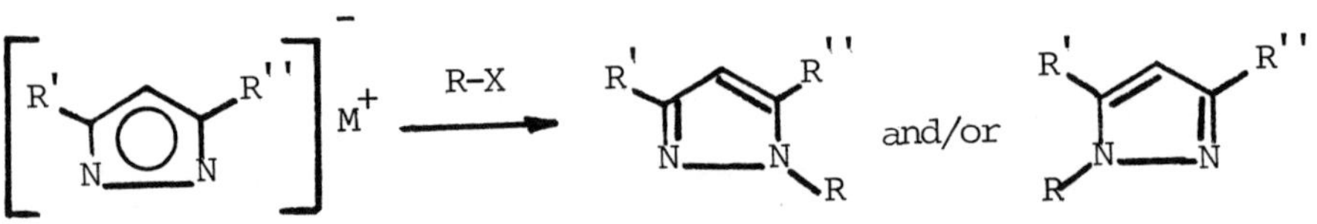

Figure 4.

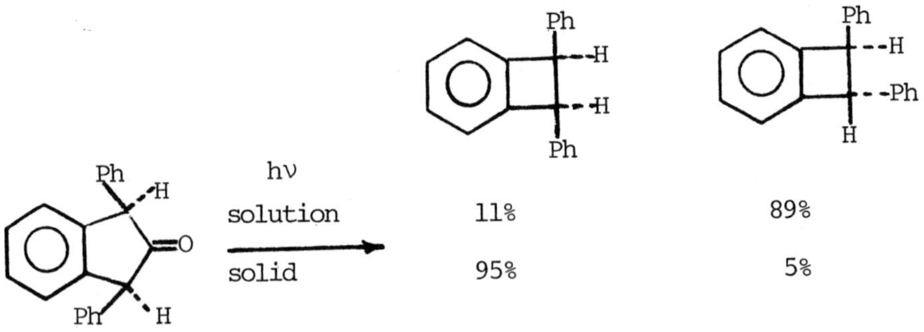

Figure 5.

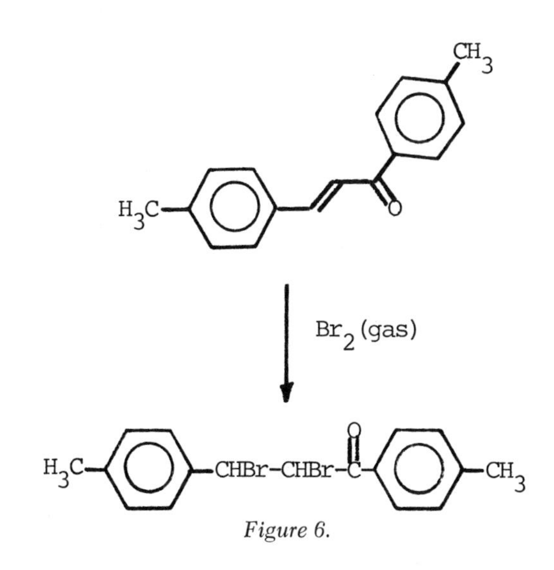

Figure 6.

$$C_6H_5\text{–}COOH + NH_3 \longrightarrow C_6H_5\text{–}COO^- \; NH_4^+$$

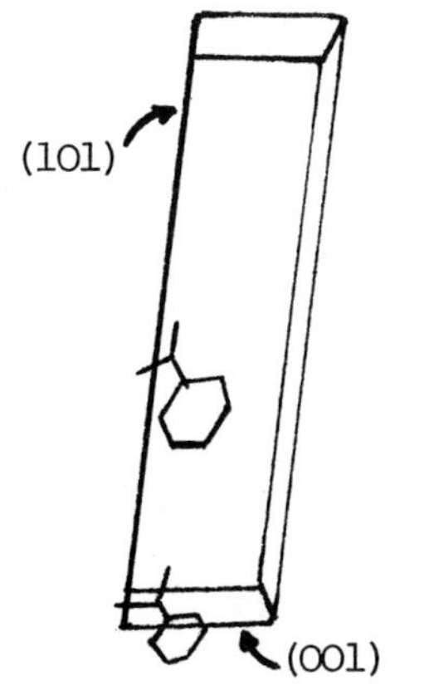

Figure 7.

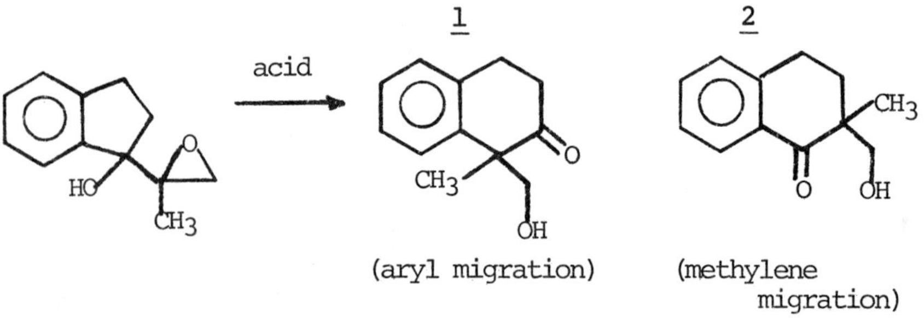

Figure 8.

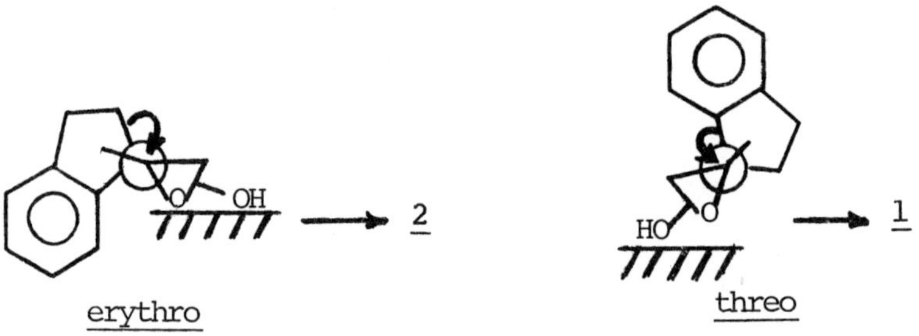

Figure 9.

Chihara (10) has recently utilized reactivity on an alumina surface in another interesting manner. Generally it is difficult to react with good yield only one of two identical functional groups in a molecule. For example, mono-esterification of terephthalic acid is complicated by diacid and diester formation. However, when terephthalic acid is chemisorbed onto alumina, then it can be mono-esterified quantitatively in a stream of diazomethane (Figure 10). The carboxyl bound to the alumina surface resists esterification, thus leading to mono-derivitization.

Lattice control of free radicals from the photolysis of crystalline acetyl benzoyl peroxide was observed by Karch and McBride (11). When acetyl benzoyl peroxide crystals are photolyzed to 15-17% completion, methyl benzoate and toluene are formed as the only detectable products (Figure 11). It was shown that the products arise exclusively from an intramolecular process in contrast to the situation with melt photolyses. Moveover, the investigators demonstrated by means of ^{18}O labelling experiments that the methyl benzoate is formed predominantly from methyl coupling with the peroxidic oxygen (as opposed to the carbonyl oxygen). Such preference for one possible oxygen over the other is absent in the liquid phase.

Our own experience with reactions on crystal surfaces has been confined to oxidations on solid $KMnO_4$ and $NaMnO_4$ (12). The work began with a curious observation in conflict with literature statements. We found that solid $KMnO_4$ stirred with a benzene solution of a secondary alcohol would oxidize the alcohol to a ketone, Regen and Koteel (13), on the other hand, reported no oxidation under these conditions. Ultimately the discrepancy was resolved with the realization that traces of water, perhaps monolayer amounts, are necessary for oxidation to take place. Slight differences in the moisture content of the $KMnO_4$ account for the varying results. When we dried the $KMnO_4$ over P_2O_5 under reduced pressure, the $KMnO_4$ lost all its oxidative ability; this could be restored by adding small quantities of moisture to the system. In addition, it was shown that $CuSO_4(H_2O)_5$ mixed with the $KMnO_4$ accelerated the oxidation of secondary alcohols. The combination produced an oxidant which was capable of oxidizing alcohols to ketones in high yield and under mild conditions as shown in Table I. Perhaps the most useful feature of the reactions in Table I is the easy workup; filtration of the solids and evaporation of the benzene give the product in a satisfactory state of purity.

Note that Table I shows that primary alcohols are not effectively oxidized by the $KMnO_4/CuSO_4$ couple. This contrasts with the usual solution behavior where primary alcohols are oxidized faster than secondary alcohols. Thus, the solid phase system presented the possibility of carrying out selective oxidations not easily accomplished in solution:

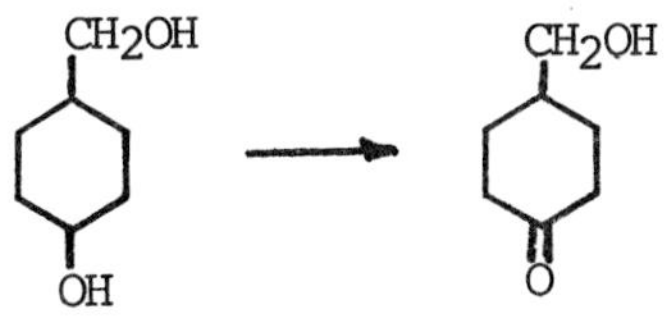

CH2OH / cyclohexane / OH → CH2OH / cyclohexane / O

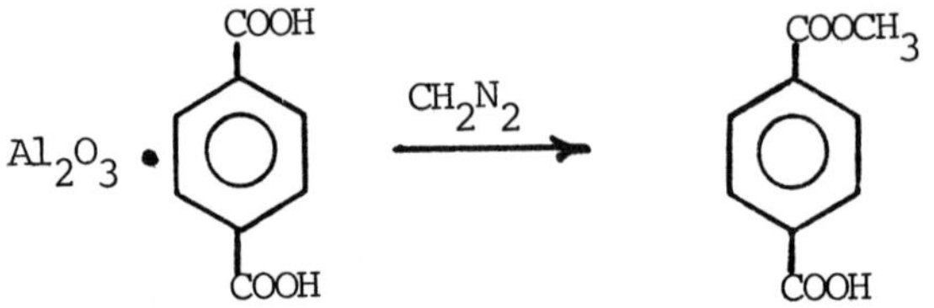

Figure 10.

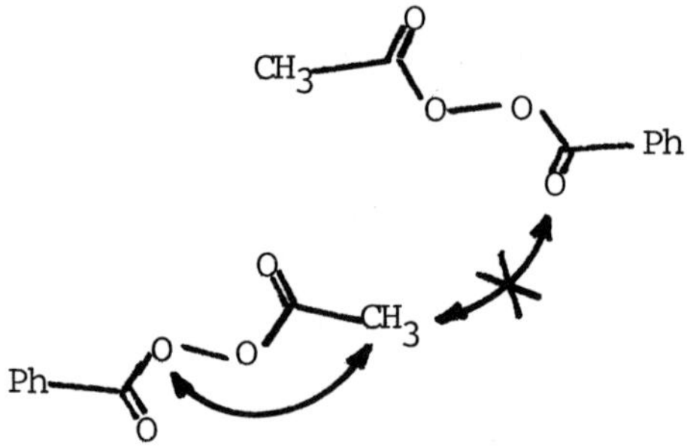

Figure 11.

Table I. Oxidations of Alcohols in Benzene[a] to the Corresponding Carbonyl Compounds by Solid Mixtures of $KMnO_4$ and $CuSO_4(H_2O)_5$

alcohol (wt, mg)	temp, °C	$KMnO_4$,g	$CuSO_4$,g	reaction time,h	yield, %[b]
2-octanol (50)	70	0.25	0.25	4	96
2-hexadecanol (2400)	70	10	10	3	100 (84)[c]
Benzhydrol (180)	70	0.50	0.50	4	100
1-cyclohexylethanol (130)	70	0.50	0.50	4	96
3-methylcyclohexanol (120)	25	0.50	0.50	2	97
ethyl lactate (120)	25	1.0	0.50	8	73
cholestanol (190)	25	1.0	0.50	11	91[c]
1-octanol (130)	25	1.0	0.50	20	<20[d]

[a]Reactions were all carried out in 3 mL of benzene except for the second entry which used 50 mL.
[b]Yields were determined by GLC using internal standards except where indicated otherwise.
[c]Isolated yield. [d]Consisted of roughly equal amounts of aldehyde and acid.

$$CH_3CH_2CH_2-\underset{\displaystyle OH}{CH}-CH_2CH_2CH_2CH_2OH \longrightarrow CH_3CH_2CH_2-\underset{\displaystyle O}{\overset{}{\underset{\|}{C}}}-CH_2CH_2CH_2CH_2OH$$

Unfortunately, such selective oxidations were not possible because primary alcohols form small amounts of carboxylic acid under the reaction conditions; these acids efficiently inhibit oxidation, thereby accounting for the apparent inertness of primary alcohols. When a secondary alcohol was mixed with a primary alcohol, neither was oxidized by the solid reagent in good yield. An oxidation of a secondary alcohol over crystalline $KMnO_4$ would immediately cease when octanoic acid was added in tiny amounts to the benzene. The octanoic acid presumably binds to "active sites" at the crystal surface and impedes the reaction by an unknown mechanism.

Sodium permanganate, commercially available although more expensive than potassium permanganate (14), was found to be an even more potent oxidant. The monohydrate form ($NaMnO_4 \cdot H_2O$) can be used as purchased to oxidize a variety of functionalities by stirring the solid with substrates dissolved in hexane or methylene chloride (Table II). The solid phase oxidation would seem particularly useful for small-scale reactions (< 1 gram substrate). Note that the $NaMnO_4 \cdot H_2O$ oxidizes olefins only slowly whereas in solution glycols are formed readily from olefins. Selectivity may turn out to be a major advantage of solid state reactions.

At present we do not understand most aspects of the solid phase oxidations: the need for trace quantities of water, Cu^{++} catalysis, RCOOH inhibition, cation dependence, reaction selectivity, crystal deterioration, intermediates and mechanism. It is a case of a field in its infancy. We will clearly need to examine spectrometrically the permanganate surface while a reaction is in progress if we are to learn the secrets of reactivity on this crystalline solid (15).

Table II. Oxidations by Solid Sodium Permanganate Monohydrate[a,b]

Substrate	Product	Solvent	Temp. (°C)	Time (h)	Yield (%)
2-octanol	2-octanone	hexane	69	2.5	95
cyclohexanol	cyclohexanone	hexane	69	1.5	100
5α-androstan-17β-01	5α-androstan-17-one	CH_2Cl_2	41	24	84
1-octanol	octanoic acid[c]	hexane	69	5	67
octyl aldehyde	octanoic acid[c]	hexane	69	4.5	77
benzyl alcohol	benzoic acid	hexane	69	6	81
benzaldehyde	benzoic acid	hexane	69	5	80
2-cyclohexen-1-ol	2-cyclohexen-1-one	hexane	69	24	47
1-octen-3-ol	1-octen-3-one	hexane	69	24	11
1-tridecene	lauric acid[c]	hexane	69	24	13
trans-stilbene	benzoic acid	CH_2Cl_2	41	24	4
	benzaldehyde				5
n-butyl sulfide	*n*-butyl sulfone	hexane	69	24	91
t-butylamine	2-methyl-2-nitropropane	hexane	69	24	76
caproamide	no reaction	CH_2Cl_2	41	24	--
1-decyne	no reaction	hexane	69	24	--
1,2-epoxytridecane	no reaction	hexane	69	24	--
n-butylbenzene	no reaction	hexane	69	24	--

[a]Reactions were carried out using 3 mmol of substrate. [b]Except for runs 2 and 8, yields are based on weights of isolated product. [c]Work-up included an extraction of the hexane with dilute $HCl/NaHSO_3$.

Literature Cited

1. Cohen, M. D.; Green, B. S., Chem. Brit., 1973, 9, 490.
2. Green, B. S.; Lahav, M.; Rabinovich, D., Acc. Chem.Res., 1979 12, 191.
3. Kornblum, N.; Lurie, A. P., J. Am. Chem. Soc., 1959, 81, 2705.
4. Taylor, E. C.; McKillop, A., Acc. Chem. Res., 1970, 3, 338.
5. Mayer-Sommer, G.; PhD thesis, Weizmann Institute of Science, 1972.
6. Quinkert, G.; Tabata, T.; Hickman, E. A. J.; Dobrat, W., Angew. Chem., Int. Ed. Engl., 1971, 10, 199.
7. Penzien, K.; Schmidt, G. M. J., Angew. Chem., Int. Ed. Engl., 1969, 8, 608.
8. Miller, R. S.; Curtin, D. Y.; Paul, I. C., J. Am. Chem. Soc, 1971, 93, 2784.
9. Cheer, C. J.; Johnson, C. R., J. Am. Chem. Soc., 1968, 90, 178.
10. Chihara, T. J., Chem. Soc. Chem. Comm, 1980, 1215.
11. Karch, N. J.; McBride, J. M., J. Am. Chem. Soc., 1972, 94, 5092.
12. Menger, F. M.; Lee, C., J. Org. Chem., 1979, 44, 3446.
13. Regen, S. L.; Koteel, V., J. Am. Chem. Soc., 1977, 99, 3837.
14. Carus Chemical Co., LaSalle, Illinois.
15. We thank the National Science Foundation, National Institutes of Health, and the Petroleum Research Foundation for support of this work.

RECEIVED June 8, 1981.

14

Photochemical Studies of Zeolites

STEVEN L. SUIB, OVIDIU G. BORDEIANU, KERRY C. McMAHON, and DIMITRIOS PSARAS

University of Connecticut, Department of Chemistry and Institute of Materials Science, Storrs, CT 06268

The following paper reviews recent results in the field of zeolite photochemistry and describes preliminary results of photochemical studies of uranyl-exchanged zeolites. The areas of water splitting by metal-exchanged zeolites and luminescence quenching are discussed as well as some of the properties of zeolites that make them excellent media for organizing chemical reactions. Some preliminary results of the photochemistry of uranyl ions exchanged into zeolites A, X, Y, mordenite and ZSM-5 are then presented. Quenching of the uranyl zeolites with isopropanol and bulk analyses of the products of this quenching are reported. Factors affecting the luminescence of uranyl ions in zeolites and their reactivity are described at the end of this paper. Results of these studies indicate that uranyl-exchanged X, Y and ZSM-5 zeolites are the most active in the photochemical conversion of isopropanol into acetone.

The use of light for the activation of inorganic ions in a variety of catalytic materials is rapidly increasing and has given a stimulus to research in photocatalysis. Materials such as micelles (1, 2), bilayers and vesicles (3, 4, 5), thin films (6, 7), monolayer assemblies (8), and clays (9) have recently been studied as organized media that have the ability to separate charged photoproducts. These materials are important as models for the photochemical processes that occur in photosynthesis (10) and for solar energy devices (11). Another group of organized media that possess attractive features for charge separation are zeolites.

Zeolite Photochemistry

Zeolites are ordered three dimensional aluminosilicates

0097-6156/82/0177-0225$05.00/0

composed of pores capable of molecular sieving. The molecular dimensions of the pores control the species that may enter or leave the zeolite. These materials have as a backbone an anionic framework that remains intact when cations are exchanged into the pores. Since there are hundreds of zeolites, a number of critical chemical parameters like molecular size, acidity and polarity can be selected by choosing a particular zeolite. There are many reviews (12, 13) and articles (14, 15) concerning zeolites to which the reader is referred for more information.

The latest review of photochemical studies of zeolites was that of Pott and Stork (16). The general principles of photoluminescence as regards inorganic ions in host lattices are discussed and will not be described here. The review of Pott and Stork concerns many different oxide catalysts such as alumina, silica and zeolites. The zeolite work mainly deals with phosphorescence of inorganic ions such as Fe^{3+}, Mn^{2+}, and Eu^{3+} in zeolites A, Y and mordenite.

Since the review of Pott and Stork there have been few reports concerning the photochemistry of ions in zeolites. Most reports involve water splitting by ion-exchanged zeolites. Jacobs and Uytterhoeven (17) reported the cleavage of water by silver exchanged Y zeolites. In these studies a two-step process first involving sunlight irradiation and secondly thermal activation yielded oxygen and hydrogen respectively. Significant amounts (0.47 mmol/g zeolite) of hydrogen were produced when silver X zeolites were used. It is believed that the system loses reversibility either due to sintering of the silver or when hydroxyl groups are removed during the thermal treatment.

Titanium(III) exchanged 3A zeolite can also split water according to Eyring and coworkers (18). Illumination with visible light causes the evolution of hydrogen as evidenced by mass spectrometry. As with the silver system described above, the titanium 3A zeolite process is not catalytic and loses reversibility. A detailed report concerning the electron paramagnetic resonance spectra of the titanium(III) 3A zeolite system has also been recently reported (19).

Europium(III) exchanged zeolites have been studied by a number of research groups. Arakawa and coworkers (20, 21) report the luminescence properties of europium(III)-exchanged zeolite Y. Emission spectra were measured under a variety of conditions and bands for europium(II) were observed after thermal treatment of the europium(III) Y zeolites. A mechanism was proposed for the thermal splitting of water which involved the cycling of europium between the two different oxidation states. Europium Mössbauer experiments (22) also show that on thermal treatment of europium-(III) zeolites that europium(II) is formed. Stucky and coworkers (23, 24) studied the phosphorescence lifetime of these europium-(III) zeolites and showed that the inverse of the lifetime (the decay constant) was linearly related to the number of water molecules surrounding the europium(III) ion in the zeolite supercages. These studies involved zeolites A, X, Y and ZSM-5.

Very recently, Lehn and coworkers (25) have used zeolite supported metal oxide catalysts for the photoinduced generation of oxygen from water. Various metal oxides like RuO_2, IrO_2 and PtO_2 were deposited on Y zeolites which were then immersed in aqueous solutions of $Ru(bpy)_3^{2+}$ and $[Co(NH_3)_4Cl]^{2+}$. The iridium and ruthenium oxide supported materials gave the best results for oxygen generation. The authors point out many important features of zeolites for studies of this type such as the easy recovery of the noble metal containing zeolite and the influence of the zeolite structure for redox activation of organic molecules.

Lunsford and coworkers (26) have prepared a $Ru(bpy)_3^{2+}$ complex in zeolite Y and studied the quenching of oxygen and water. The emission bands of the ion-exchanged zeolite resemble those of aqueous solutions. Diffuse reflectance spectroscopy and ESCA measurements were also made in the characterization of these samples.

Faulkner and coworkers (27, 28), have also studied the interaction of $Ru(bpy)_3^{2+}$ with zeolite X. Luminescence lifetime measurements and emission spectra were used to study electron transfer quenching of the electron donors N,N,N',N'-tetramethyl-p-phenylenediamine and 10-phenyl-phenothiazine. Lifetime measurements show at least two modes of quenching for the interaction of $Ru(bpy)_3^{2+}$ ions with these donors. Products of these electron transfer reactions were isolated and these experiments show that the zeolite can separate the products of light induced electron transfer.

Uranyl-exchanged Zeolites. The same properties of a zeolite that are useful in preventing the back reaction of photoproducts are also important in a zeolite photocatalytic system. Our goal is to explore the dynamics of photolyzed uranyl-exchanged zeolites. The only report involving catalysis by a uranyl zeolite reveals that this material is productive in gas oil hydrocracking as well as toluene disproportionation (29). Since a large number of systems containing the uranyl cation have long been known to be photosensitive and catalytic (30) it is quite surprising that photochemical studies of uranyl zeolites have not been explored. Since photochemical reactions and zeolite reactions are well known to be selective, perhaps by combining the two areas a highly selective process may be developed. The luminescence, spectroscopic and crystallographic properties of uranyl-exchanged zeolites A, X, Y, mordenite and ZSM-5 are the subject of the rest of this paper.

Experimental Procedure

Ion-exchange of the zeolites was carried out in our work by suspending 1 gram of zeolite in 100 ml of 0.01M $UO_2(CH_3COO)_2 \cdot 2H_2O$ and stirring the mixture for 24 hours. After the ion-exchange the zeolites were filtered, washed with pure distilled deionized water and dried *in vacuo* at room temperature. The details of this procedure are available elsewhere (31).

Luminescence spectra were recorded on a double Czerny-Turner scanning monochromator Model 1902 Fluorolog Spex spectrofluorometer. Variations in the excitation radiation are automatically corrected by a reference detector equipped with a Rhodamine B quantum counter. Emission spectra were recorded with the right angle mode. Further details are available elsewhere (31).

Bulk photolyses were carried out using a 1000-W, high pressure, Xe lamp (Model 6117, Oriel Corp., Stamford, Conn. 06902) and a UV-VIS grating monochromator. For purposes of comparison and evaluation of power, 8.5 mW of power at 425 nm is produced at the sample surface within the cuvette cell. The procedure for determining this is found elsewhere (32). Samples were loaded in a type 52-H 2mm light path quartz cell purchased from Precision Cells, Inc., Hicksville, NY. The amount of zeolite used was 0.3 grams. Solutions of 1.5 M isopropanol dissolved in acetonitrile were used for the bulk photolyses.

Gas chromatographic analyses were performed on a Hewlett Packard Model 5880 system equipped with a thermal conductivity detector with an injector temperature of 210°C, detector temperature of 210°C and an oven temperature set at 60°C. All runs were isothermal. The column used for the separation was a 6 foot 5% carbowax 20 M, 80-100 mesh, W.A.W. DMCS. Absolute retention times were 1.11 min. (acetone), 1.95 min. (isopropanol) and 3.5 min. (acetonitrile). Methods of automatic peak integration that were used during the analyses were peak height, peak area, area % and internal standard experiments.

Analyses of the zeolites after ion-exchange were made with a Diano-XRD 8000 X-ray powder diffraction apparatus.

Results

Emission Spectra. The emission spectra of the uranyl acetate dihydrate in solution and in the solid state are shown in Figure 1. The fine structure in the solid state spectrum is not observed in solution. The corresponding emission spectra of uranyl-exchanged zeolites, A, Y, mordenite and ZSM-5 are shown in Figures 2-4. Excitation is carried out at 366 nm. The emission spectra have been scanned in all cases between 450 nm and at least 630 nm.

Further observations from similar experiments are that the emission spectral lineshapes are not a function of the uranyl ion concentration in the zeolite. The lineshape does also not seem to be influenced by the relative acidity of the ion-exchanged zeolites. When the degree of hydration is changed from fully hydrated (stored in a dessicator under saturated NH_4Cl aqueous solutions) to vacuum dried at 1×10^{-3} Torr at room temperature, only the intensity of luminescence (and not the lineshape) changes.

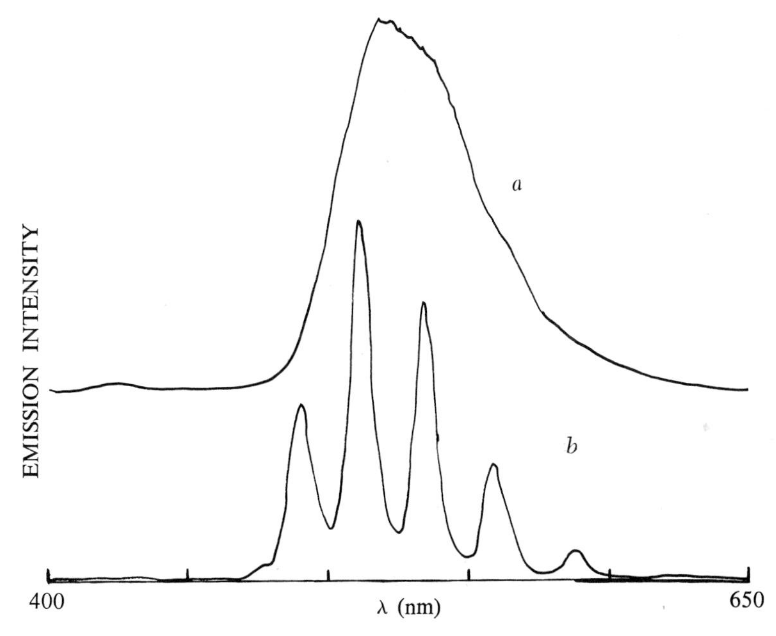

Figure 1. a, emission spectrum of $UO_2(CH_3COO)_2 \cdot 2\ H_2O$ *in aqueous* 10^{-2} M *solution; and b, emission spectrum of* $UO_2(CH_3COO)_2 \cdot 2\ H_2O$ *in the solid state. Excitation was carried out at 366 nm.*

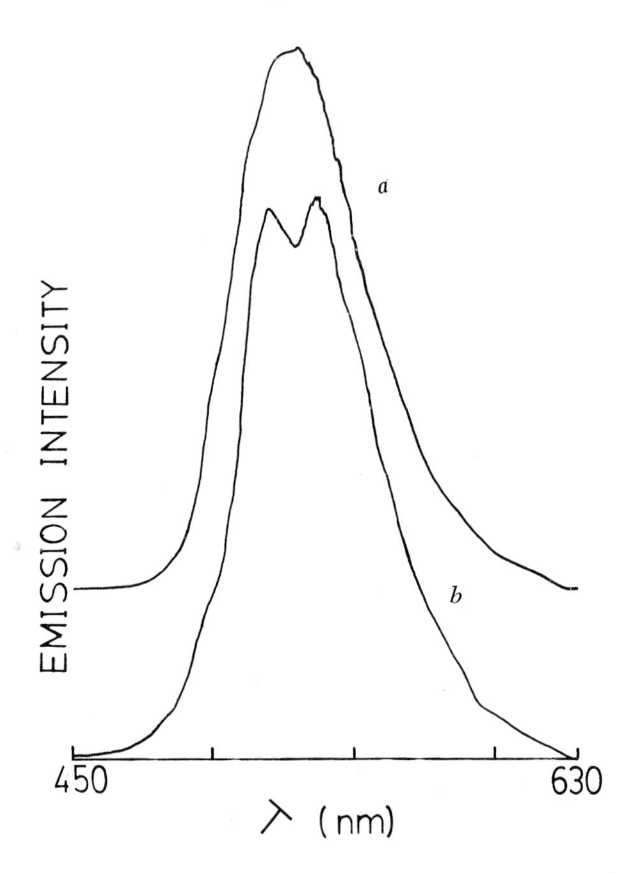

Figure 2. a, emission spectrum of UO_2^{2+} exchanged A zeolite; and b, emission spectrum of UO_2^{2+} exchanged ZSM-5 zeolite. Excitation was carried out at 366 nm.

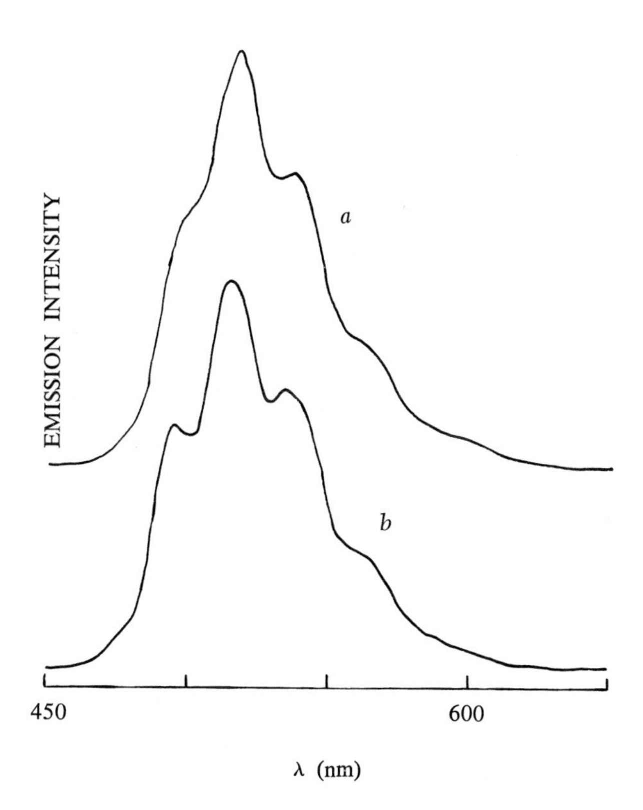

Figure 3. a, emission spectrum of UO_2^{2+} exchanged NH_4Y zeolite; and b, emission spectrum of UO_2^{2+} exchanged NaY zeolite. Excitation was carried out at 366 nm.

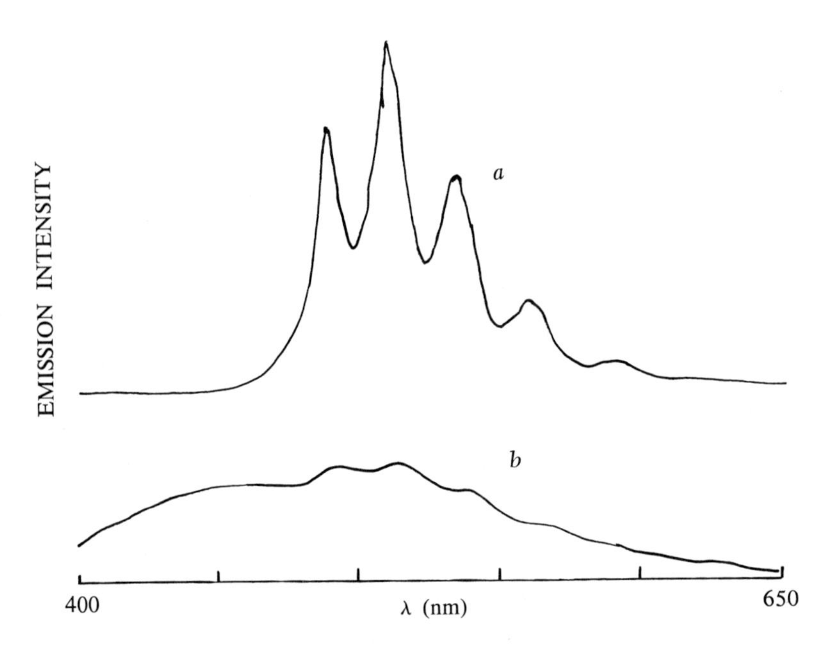

Figure 4. a, emission spectrum of UO_2^{2+} exchanged Na mordenite; and b, emission spectrum of UO_2^{2+} exchanged H mordenite. Excitation was carried out at 366 nm.

A Stern Volmer plot of the quenching of uranyl-exchanged Y zeolite by isopropanol is given in Figure 5. The I_0/I value was determined by measuring the peak intensity without any quencher (I_0) and dividing this by the peak intensity with quencher (I). This ratio was determined by taking the value of the maximum intensity (∿522 nm), although I_0/I ratios were consistent over a wide wavelength range (500 nm to 550 nm).

Bulk Photolyses. Data for bulk photolyses of uranyl-exchanged zeolites in contact with isopropanol/acetonitrile mixtures are given in Table I. The X-ray powder diffraction data reveal that the UO_2A sample and the UO_2ZSM-5 sample are both amorphous. All other samples are crystalline after the ion exchange procedure. Crystallinity is not lost when the samples are photolyzed.

Discussion

Ion-exchange of uranyl ions into zeolites has been previously described (33, 34). The uranyl ion is certainly small enough to exchange into zeolites X, Y and mordenite. The X-ray powder diffraction data reveal that the exchanged A zeolite and the exchanged ZSM-5 zeolite are amorphous. In the case of zeolite ZSM-5 the pores are still filled with the tetrapropyl ammonium polymeric network used as a backbone or template around which the zeolite crystallizes (35). Perhaps because of this fact the uranyl ions can not be exchanged into the interior of the zeolite. If the starting unexchanged ZSM-5 zeolite is thermally treated, however, the polymeric material is removed and exchange can take place readily. The amorphous nature of the uranyl-exchanged zeolites A and ZSM-5 correlates well with their luminescence emission spectra which do not have much fine structure and resemble the solution spectrum of uranyl acetate dihydrate given in Figure 1. Collisional deactivation in solution accounts for this loss in vibrational fine structure.

The emission spectra for uranyl-exchanged zeolites Y, mordenite and X all have differences but do show some fine structure and therefore resemble the solid state spectrum of uranyl acetate dihydrate. In fact, the spectrum of uranyl ions exchanged into sodium mordenite is very similar to that of the uranyl acetate dihydrate solid spectrum shown in Figure 1. Further support for our belief that some zeolites have a solution like environment and others have a solid like environment comes from the correlation between the crystallinity of these uranyl-exchanged zeolites and the appearance of some fine structure in the emission spectrum. We find no apparent correlation between this fine structure and the concentration of the uranyl ion in the zeolites even with a ten-fold change in the concentration of the uranyl ion. The degree of hydration also does not seem to change the emission spectrum unless the ion-exchanged material is heated. Preliminary

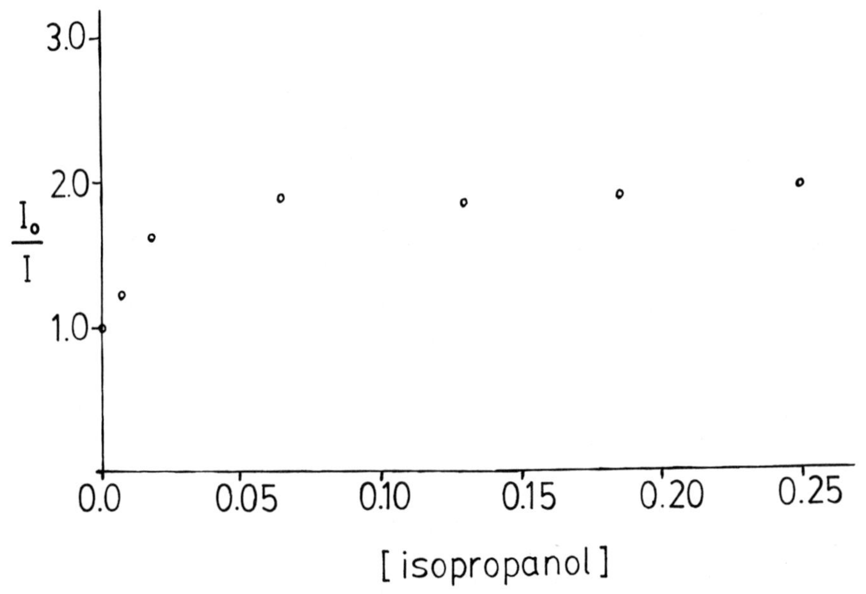

Figure 5. Stern-Volmer plot for quenching of UO_2^{2+} exchanged Y zeolite by 2-propanol.

Table I.

Bulk Photolysis of Isopropanol Uranyl Zeolite Suspensions

Zeolite[a]	Photolysis Time[b]	[Isopropanol][c]	Moles Acetone[d]
UO_2A	30	1.05	e
UO_2mordenite	30	1.05	e
UO_2mordenite(H^+)	30	1.05	e
UO_2ZSM-5	30	1.05	2.14
UO_2X	30	1.05	0.90
UO_2Y	10	1.05	0.20
UO_2Y	30	1.05	0.46
UO_2Y	45	1.05	1.06

a. 0.30 grams of zeolite used.
b. 1000 Watt Xe lamp, 425 nm wavelength, 8.5 mW of power, time in minutes.
c. Acetonitrile as solvent.
d. Carbowax 20-M column used, moles acetone times 10^{-5}.
e. No acetone detected.

acidity measurements of the uranyl-exchanged zeolites also reveal no correlation between the acidity and the fine structure.

The most reasonable qualitative explanation of the change in the emission spectra is the above-mentioned resemblance to solution or solid state behavior which correlates well with the crystallinity of the uranyl-exchanged zeolites as determined by X-ray powder diffraction. However, there are differences between all of the zeolites which could be an indication of site symmetry and coordination in the lattice. The aluminosilicate lattice is preserved after ion-exhange except for zeolites A and ZSM-5.

Isopropanol Quenching and Acetone Generation. The Stern Volmer plot of Figure 5 shows a dual quenching pattern similar to plots observed in the quenching of $Ru(bpy)_3^{2+}$ zeolite X materials with N,N,N',N'-tetramethyl-p-phenylenediamine and 10-phenyl-phenothiazine (27, 28). For those experiments it is believed that the dual quenching is due to two different quenching sites, one being external and the other internal, with respect to the zeolite lattice.

Further support for this explanation is given by the uranyl-exchanged Y zeolite Stern Volmer plot shown in Figure 5 as well as the data in Table I. It is observed from the data in Table I that large pore zeolites like zeolites X and Y tend to produce significant amounts of acetone on bulk photolysis. There is also an induction time of about 10 minutes before which acetone is not detected by gas chromatographic techniques. These observations are in line with a fast relatively small quenching of isopropanol by external surface uranyl ions which does not produce large amounts of acetone (for instance, zeolite A) followed by diffusion into the supercages and quenching by internal uranyl ions which are much more abundant than the external surface uranyl ions. The induction time could be due to diffusion into the supercages of the zeolites. It is also noted that the conditions for this photolysis are quite mild, with very short photolysis times producing relatively large amounts of acetone. The excitation is in the visible region and no other products from any of the photolyses are detected. This helps support our hopes that these zeolite photolysis reactions can be highly selective.

Future Areas of Research. There are certain future experiments that are very important for further understanding of the work reported here. The most important ones concern lifetime measurements of the quenching of isopropanol. Secondly, the mechanism of quenching must be understood. Electron paramagnetic resonance experiments will be helpful here. More bulk photolyses and other organic quenchers need to be studied, especially variable size quenchers, in order to help understand diffusional processes in these zeolites. We are presently studying the surfaces of these zeolites with ion scattering spectrometry and secondary ion mass spectrometry. The preliminary results indicate that we may be able to distinguish external sites from internal sites in these uranyl-exchanged zeolites. The fundamental observations made in this preliminary report provide further evidence for the behavior of solid zeolites as solutions as has been recently proposed (36, 37, 38). The fate of the UO_2^{2+} cations in the zeolite after photolysis is unknown at present. Photolysis experiments in the presence and absence of O_2 are in progress.

Acknowledgements

We are grateful to the Atlantic Richfield Foundation of the Research Corporation and the University of Connecticut Research Foundation for supporting this work.

Literature Cited

1. Thomas, J. K.; Piciulo, P. in Wrighton, M. S., Ed. ACS Adv. Chem. Ser. 1980, 184.
2. Grätzel, M. Ber. Bunsenges. Phys. Chem. 1980, 84, 981.
3. Calvin, M. Accounts Chem. Res. 1978, 11, 365.
4. Grimaldi, J. J.; Boileu, S.; Lehn, J. M. Nature 1977, 265, 229.
5. Whitten, D. G. Accounts Chem. Res. 1980, 13, 82.
6. Faulkner, L. R.; Tachikawa, H.; Fan, F.-R.; Fischer, S. G. in Wrighton, M. S. Ed. op. cit.
7. Renschler, C. L.; Faulkner, L. R. Faraday Discuss. Chem. Soc. 1980, 70, in press.
8. Whitten, D. G.; Mercer-Smith, J. A.; Schmehl, R. H.; Worsham, R. R. in Wrighton, M. S., Ed. op. cit.
9. Krenske, D.; Abdo, S.; Van Damme, H.; Fripiat, J. J. J. Phys. Chem. 1980, 84, 2447.
10. Knox, R. S. in "Bioenergetics of Photosynthesis" Govindjee, Ed.; Academic, New York, 1975.
11. Wrighton, M. S. Chem. Eng. News 1980, 57 #36, 29.
12. Smith, J. V. in "Zeolite Chemistry and Catalysis" Rabo, J. A. Ed.; ACS Monograph Series, Vol. 171. American Chemical Society: Washington, 1976.
13. Eberly, P. E. Jr. in "Zeolite Chemistry and Catalysis" Rabo, J. A., Ed.; ACS Monograph Series, Vol. 171. American Chemical Society: Washington, 1976.
14. Haynes, H. W. Jr. Catal. Rev. 1978, 17, 273.
15. Rollmann, L. D. J. Catal. 1977, 47, 113.
16. Pott, G. T.; Stork, W. H. J. Catal. Rev. 1975, 12, 163.
17. Jacobs, P. A.; Uytterhoeven, J. B.; Beyer, H. K. J. Chem. Soc. Chem. Commun. 1977, 128.
18. Kuznicki, S. M.; Eyring, E. M. J. Am. Chem. Soc. 1978, 100, 6790.
19. Kuznicki, S. M.; DeVries, K. L.; Eyring, E. M. J. Phys. Chem. 1980, 84, 535.

20. Arakawa, T.; Takata, T.; Adachi, G. Y.; Shiokawa, J. J. Lum. 1979, 20, 325.
21. Arakawa, T.; Takata, T.; Adachi, G. Y.; Shiokawa, J. J. Chem. Soc. Chem. Commun. 1979, 453.
22. Suib, S. L.; Zerger, R. P.; Stucky, G. D. Emberson, R. M.; Debrunner, R. G.; Iton, L. E. Inorg. Chem. 1980, 19, 1858.
23. Zerger, R. P.; Suib, S. L.; Stucky, G. D., Abstracts of the 178th American Chemical Society National Meeting, Washington, D. C., September 1979.
24. Zerger, R. P.; Suib, S. L.; Stucky, G. D. J. Am. Chem. Soc. submitted.
25. Lehn, J. M.; Sauvage, J. P.; Ziessel, R. Nouv. J. Chim. 1980, 4, 355.
26. DeWilde, W.; Peeters, G.; Lunsford, J. H. J. Phys. Chem. 1980, 84, 2306.
27. Faulkner, L. R.; Suib, S. L.; Renschler, C. L.; Green, J. M.; Bross, P. R. in "Chemistry in Energy Production", Wymer, R. G.; Keller, O. L., Eds., 1981, in press.
28. Suib, S. F.; Renschler, C. L.; Green, J. M.; Bross, P. R.; Faulkner, L. R., submitted.
29. Bertolacini, R.; Gutberlet, L. C. U. S. Patent 3,650,945, 1972.
30. Rabinowitch, E.; Belford, R. L. "Spectroscopy and Photochemistry of Uranyl Compounds"; Pergamon, Oxford, 1964.
31. Bordeianu, O. G.; McMahon, K. C.; Psaras, D.; Suib, S. L. 1981, submitted.
32. Eaton, H. E.; Stuart, J. D. Anal. Chem. 1978, 50, 587.
33. Mocevar, S.; Drzay, B.; Zajc, A. J. Inorg. Nucl. Chem. 1979, 41, 91.
34. Onu, P.; Ababi, V. Rev. Roun. Chim. 1974, 19, 1279.
35. Argauer, R. J.; Landolt, G. R. U. S. Patent 3,702,886, 1972.
36. Barthomeuf, D. J. Phys. Chem. 1979, 83, 249.
37. Morrison, T. I.; Iton, L. E.; Shenoy, G. K.; Stucky, G. D.; Suib, S. L.; Reis, A. H. J. Chem. Phys. 1980, 73, 4705.
38. Chem. Eng. News 1981, 59 #15, 32.

RECEIVED May 7, 1981.

INDEX

INDEX

H

I

Q

T

Y

Z

Jacket design by Kathleen Schaner.
Production by Katharine Mintel and V. J. DeVeaux.

Elements typeset by Service Composition Co., Baltimore, MD.
Printed and bound by The Maple Press Co., York, PA.

S0-CCA-495
我亲爱的艾丽斯，
等你到来的日子里，我有了一 伙
伴，一只海鸥。我猜它折断 翅膀，
别担心，它看起来在慢慢恢 ，我
希望我们的灯塔运转良好 更想念
陪伴.

献给苏珊·里奇。

——我的编辑、朋友，她就像是我人生灯塔上的明灯。

图书在版编目（CIP）数据

你好灯塔 /（澳）苏菲·布莱科尔著绘；范晓星译
. -- 北京：中信出版社，2019.4
书名原文：Hello Lighthouse
ISBN 978-7-5217-0088-6

Ⅰ．①你… Ⅱ．①苏… ②范… Ⅲ．①儿童故事－图画故事－澳大利亚－现代 Ⅳ．①I611.85

中国版本图书馆 CIP 数据核字 (2019) 第 028946 号

你好灯塔

著 绘 者：[澳] 苏菲·布莱科尔
译　　者：范晓星
出版发行：中信出版集团股份有限公司
（北京市朝阳区惠新东街甲 4 号富盛大厦 2 座　邮编　100029）
承 印 者：深圳当纳利印刷有限公司

开　　本：889mm × 1194mm 1/12　　印　　张：4.5　　字　　数：60 千字
版　　次：2019 年 4 月第 1 版　　印　　次：2019 年 4 月第 1 次印刷
京权图字：01-2019-0995　　广告经营许可证：京朝工商广字第 8087 号
书　　号：ISBN 978-7-5217-0088-6
定　　价：49.80 元

你好灯塔

[澳] 苏菲·布莱科尔 著绘　范晓星 译

中信出版集团 | 北京

在遥远的天边有一座很小的岛，
岛上礁石的最高处矗立着一座灯塔。
它恒久地站立在那里，
面向大海，放射光芒，
为过往的航船指引方向。

从黄昏到黎明，灯塔放射出一束束光芒。

你好！

你好！

你好！

你好，灯塔！

新来的守塔人接替年迈的守塔人，
继续守护塔灯。
他擦亮透镜，注满灯油，
修剪燃过的灯芯。
夜晚，他按时上好发条，
保证塔灯旋转。
白天，他为圆形的房间刷上新油漆，
如海水般深绿的油漆。
他记下灯塔日志，穿针引线，
聆听四面八方的风声。

风深深地吸了一口气，吹呀，吹呀。
你好！
你好！
你好！

守塔人烧好水，为自己沏上一杯茶，
然后探身窗外钓鳕鱼。
他哼着小曲，摆好餐具，
多希望能有个人说说话。

每隔几天，他给她写一封信，
让海浪带走。
他守护塔灯，记下灯塔日志，
等候她的回信。

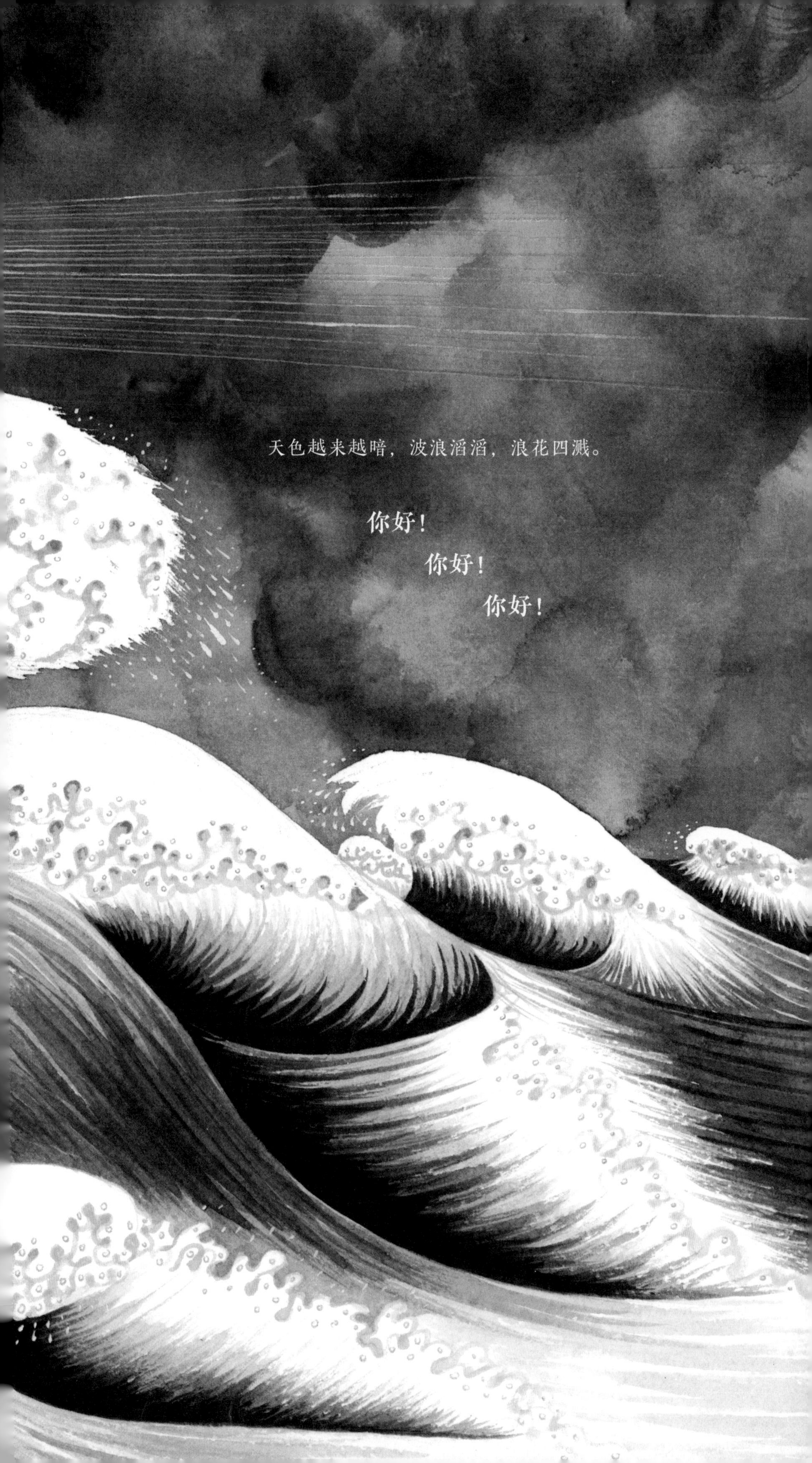

天色越来越暗，波浪滔滔，浪花四溅。

你好！

你好！

你好！

守塔人用望远镜眺望远方。

灯塔勤务船来了，送来了灯油、面粉、猪肉、豆子……

和他的妻子。

他领着妻子在灯塔里转了一大圈儿，
参观了里面每个圆形的房间。
他守护塔灯，
记下灯塔日志，
摆上两套餐具。

一切都消失在浓雾之中，

必须敲响雾钟，提醒过往的航船！

铛！

铛！

铛！

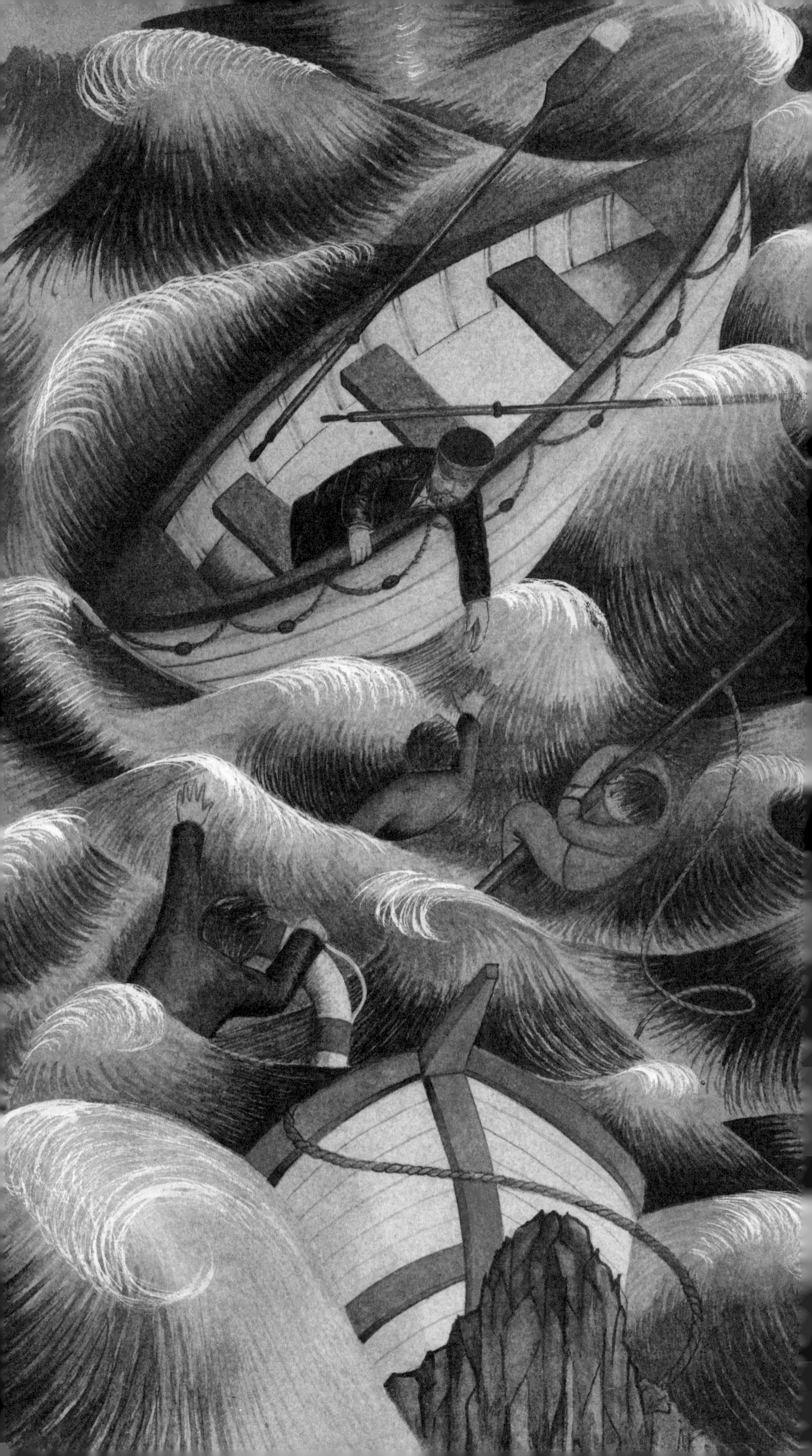

在伸手不见五指的黑夜，海难发生了！
一艘小船撞上了礁石！
一刻都不能耽误，守塔人划着木船出发了。
他将三名水手从深不可测、漆黑如墨的海水里拉上船。
他守护塔灯，记下灯塔日志，
为水手们披上毯子。

海面结了一层厚厚的冰。

你好！

你好！

你好！

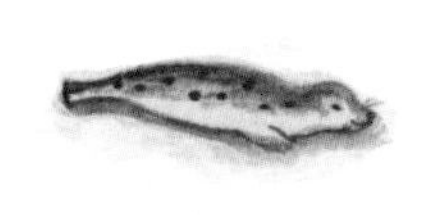

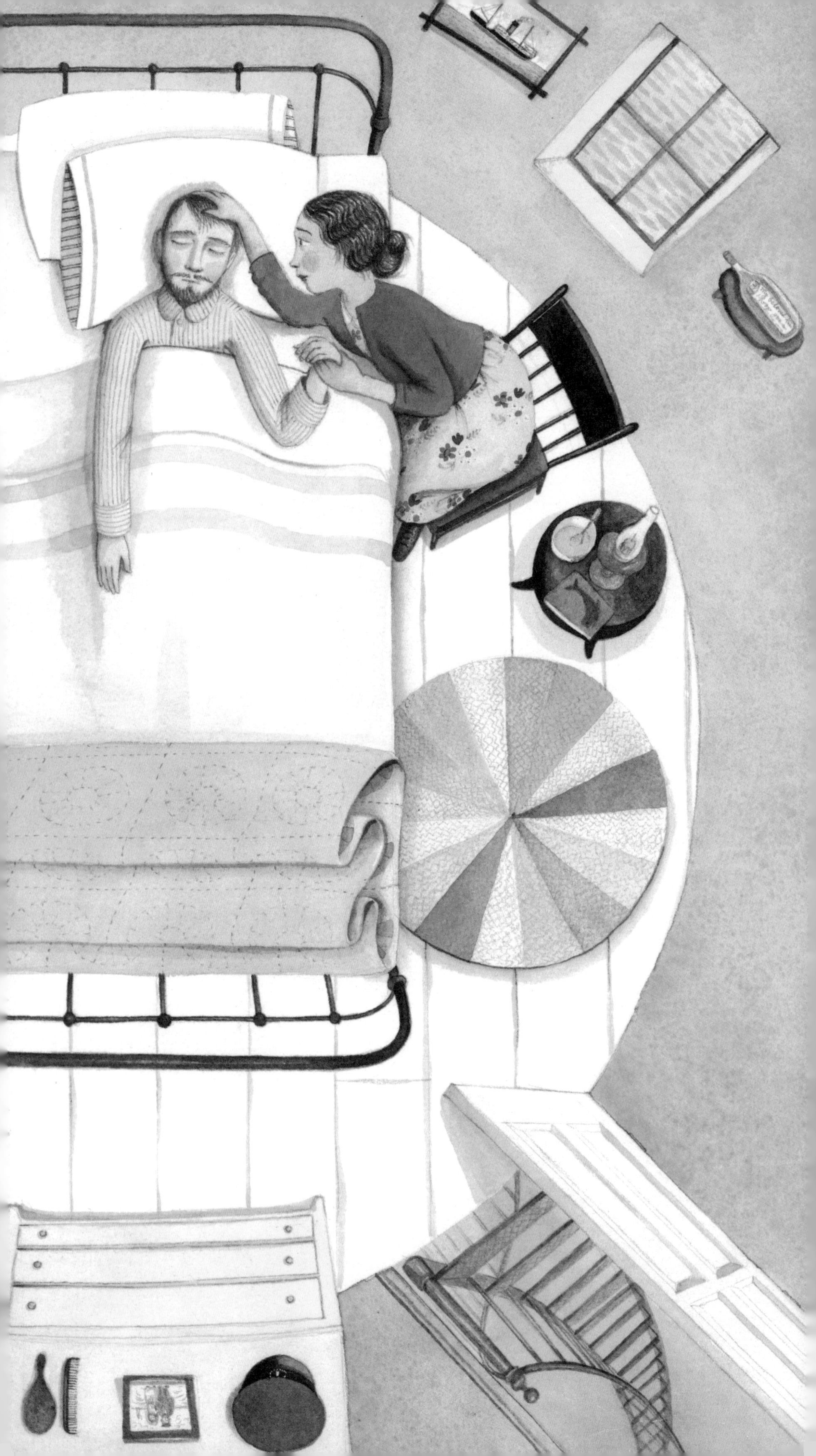

一天清晨，守塔人打起了喷嚏；
黄昏时分，他已病得下不了床。

他的妻子顿时

里里外外忙个不停，

沿着旋转楼梯跑上跑下。

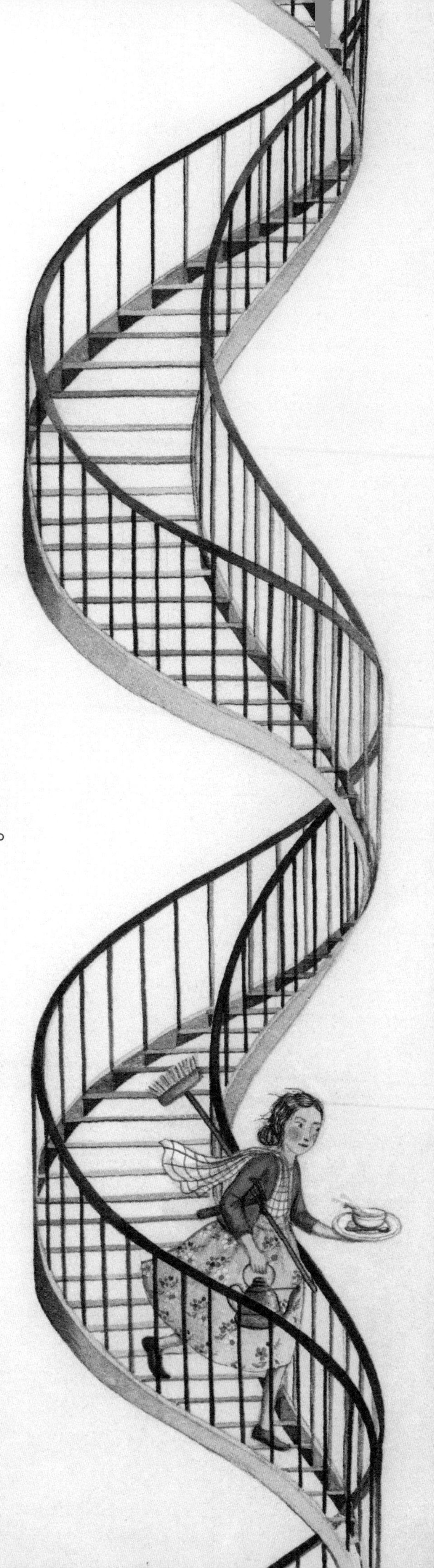

她守护塔灯，
给丈夫喂汤，
铲除塔灯室窗户上的冰霜。
她坐在丈夫身边，
在灯塔日志上记录下他退烧的那一刻。

冰山向着南方漂去。
鲸鱼朝着北方遨游。

你好！

你好！

你好！

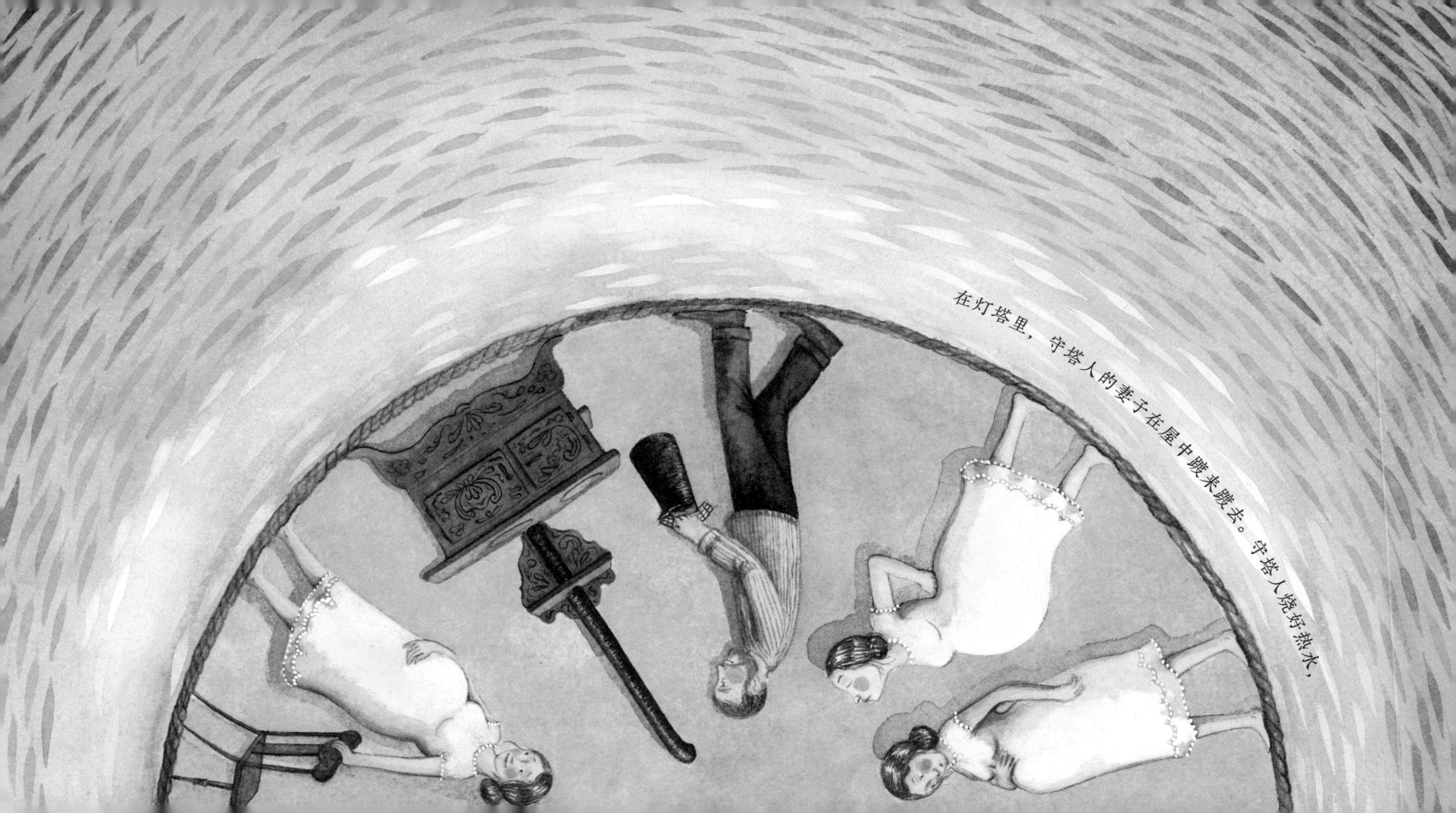
在灯塔里，守塔人的妻子在屋中踱来踱去。守塔人烧好热水，

搀扶着她，吸气，呼气。他守护塔灯，在灯塔日志里……

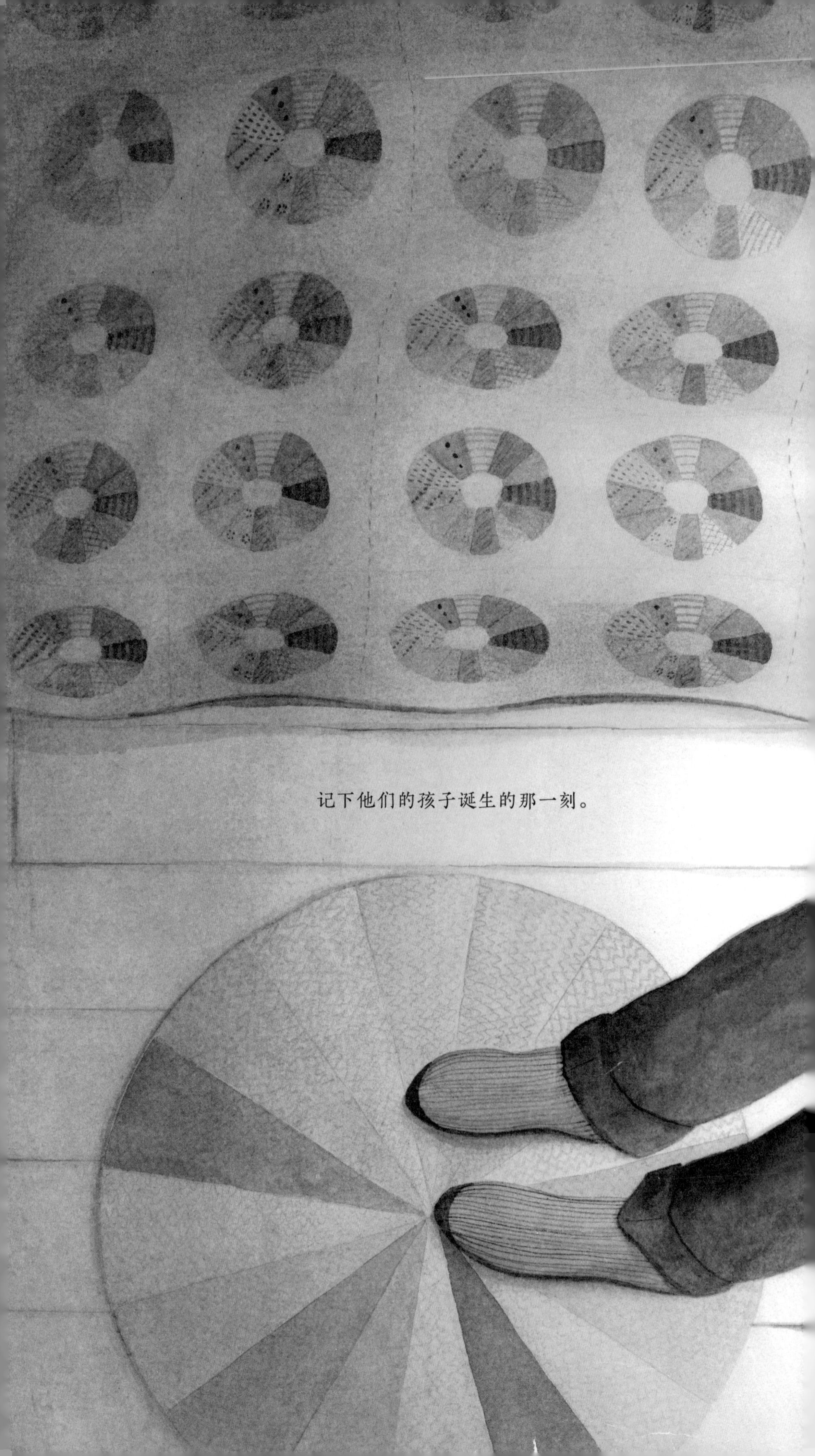

记下他们的孩子诞生的那一刻。

JOURNAL of Lighthouse
DAY
WEATHER
Whale Rock
REMARKS

天空闪现绚丽曼妙的绿光。
你好！
你好！
你好！

勤务船到了，它运来了灯油、
面粉、猪肉、豆子和邮件。

跟随新书和大陆上的消息一起来的
还有一封不期而至的信，
信封上有海岸警卫署的封蜡。
守塔人上好发条，擦亮透镜，
如往常一样。
他守护塔灯，记下灯塔日志，
可是他知道，这样的日子不长了。

一家人眺望遥远的海平线。

海岸警卫署的人来了，
他们带来了新式塔灯，
安装好运转塔灯的机器。

不用再注满灯油，
不用再修剪灯芯，
守塔人的工作结束了。

他登上旋转楼梯的最高处，
合上了灯塔日志，
永远地。

一家人整理好行装，
放到小船上，
与天上的海鸥
挥手告别。

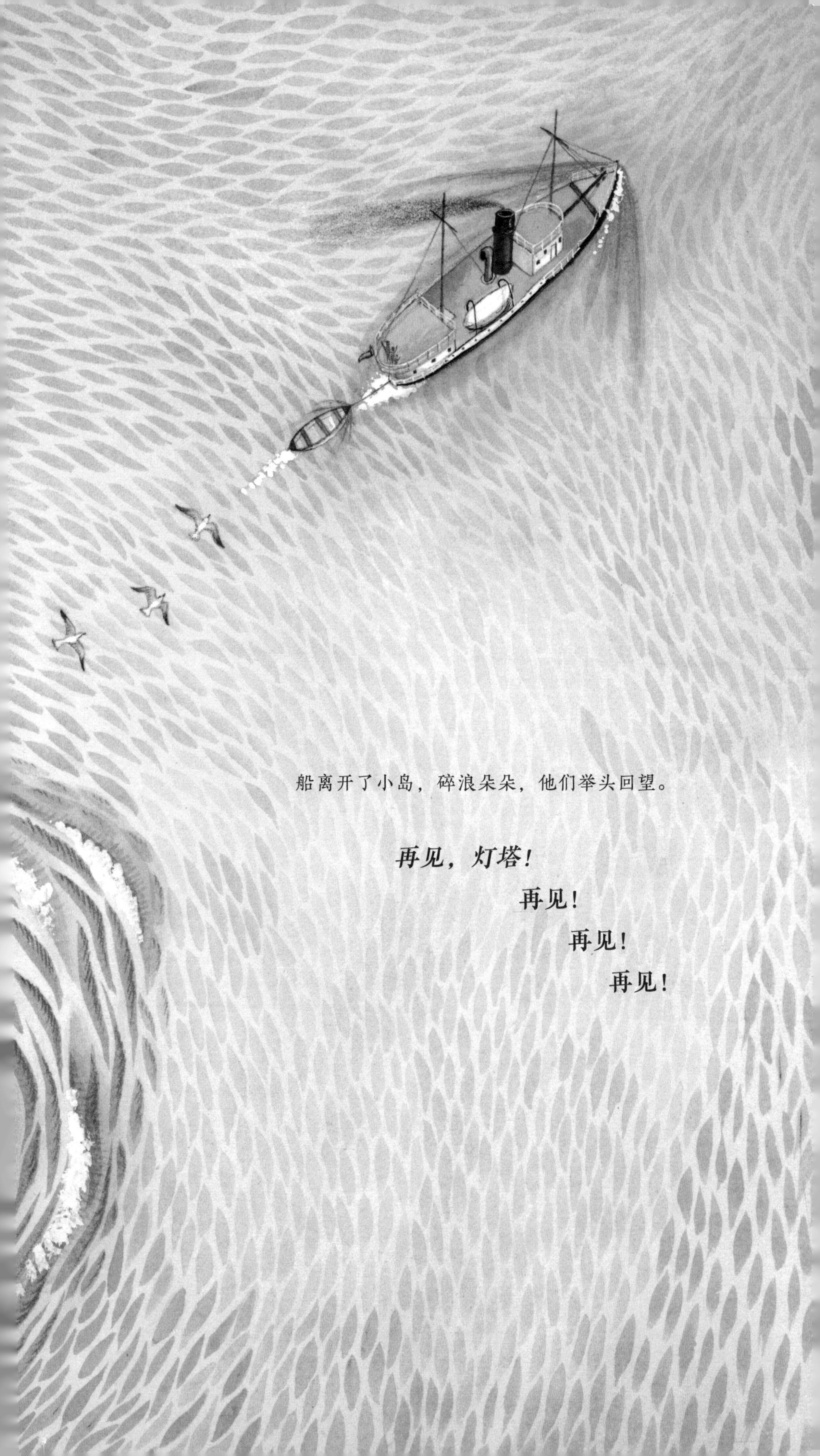

船离开了小岛，碎浪朵朵，他们举头回望。

再见，灯塔！

再见！

再见！

再见！

在遥远的天边有一座很小的岛，
岛上礁石的最高处矗立着一座灯塔。
它恒久地站立在那里，
面向大海，放射光芒。

雾起，雾散。
浪起，浪落。
风，吹来，吹去。

你好！

你好！

你好吗？

越过海浪，跨过海湾，陆地上有一束光芒，它仿佛在回答：

你好！

你好！

你好！

你好，灯塔！

My beloved Alice

作者的话

苏菲·布莱科尔

以前我都是从航船的角度来认知灯塔的：它像一位忠于职守的哨兵，傲然矗立，向着大海放射光芒，提醒船只哪里有危险的礁石，引导水手安全航行。但是有一天，我在跳蚤市场偶然见到一张老画片，这张灯塔内部剖面图映入我的眼帘，令我产生无限遐想：如果住在灯塔里会是什么样子？生活在狭小的圆形房间中，以塔为家，远离陆地，孤独一人。后来，我展开了深入的研究，查找图片，阅读资料，参观博物馆，登过从美国纽约州向北至加拿大纽芬兰省的众多灯塔。

我了解到，守塔人的职责是从日落到日出守护塔灯，永远不能让塔灯熄灭。有时候守塔人会有一个助手，他们一起下跳棋，分担守夜的工作。但这两个人不总是合得来。据说，某个守塔人与他的属下互相厌烦，甚至一整年都不说一句话，就连吃饭的时候也是背对背，谁也不看谁。通常，守护灯塔的是一家人。据记载，有个守塔人家中一共有十一个孩子。守塔人行列中也曾有成百上千位女性。她们有的是接替父亲或者丈夫的工作，但也有很多是直接被赋予了这个使命。

夜晚，守塔人要多次起来给旋转透镜上好发条。他们要剪掉燃烧过的灯芯，给油灯注满灯油。他们要到地窖，用木桶从油罐里接满油，再提着沉重的木桶走上长长的旋转楼梯，一直提到塔灯室。他们还必须在每天早上擦去透镜上的油烟，使透镜一尘不染，这样才能保证它发出的光最明亮。在阳光明媚的日子，他们要拉上窗帘，以防透镜聚光起火。塔灯室的窗户要光洁如新。在冬天，这就意味着守塔人要刮掉窗户外面的冰霜。这是一项非常危险的工作，守塔人要牢牢抓住灯塔外面的特殊把手，有时还需要将自己绑在护栏上，以免被强风吹走。

在浓雾弥漫的天气里，船上的人无法看到塔灯，守塔人就要敲响警钟。在很早以前，守塔人要用手握锤，每隔20秒钟击钟一次。这是件让人精疲力竭的工作，更别提还得忍受震耳欲聋的钟声了。时过境迁，有发条装置的钟代替了人工敲击的钟，大型灯塔里还用上了蒸汽雾号。有些守塔

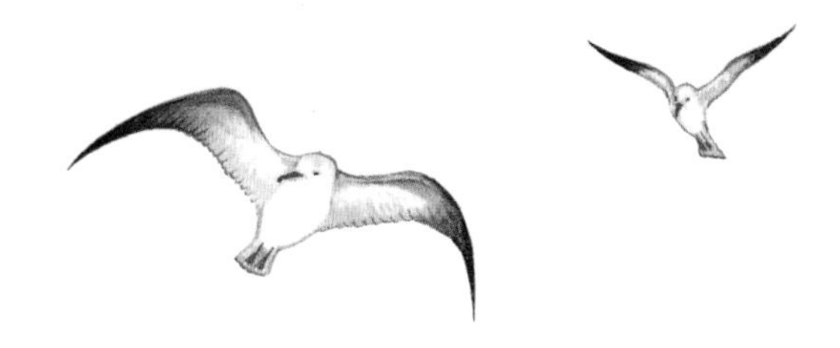

人即使在持续不断的雾号声中也能睡着；如果大雾数日不散，有些守塔人则会因此而抓狂。

因为有海水和大风的不断侵蚀，灯塔塔身需要没完没了的维护，守塔人要经常粉刷它。除此以外，当然还有烦琐的日常事务：洗衣、做饭、缝补衣服、给煤炉加煤。守塔人用的水需要手动从大水罐中压出来。他们还要把厨房里的泔水以及马桶里的污物倒进海里。

有关灯塔的所有事情都要事无巨细地记录到当天的灯塔日志里：琐事如天气、灯油用量、过往船只的种类与名称，以及一些意义非凡的大事，比如新生命降临、亲人故去或者海上事故。

即便做完了所有这些日常工作和杂务，守塔人还是会有很多的时间要打发。他们会读书，玩单人纸牌游戏，编织，刺绣，制作船模或者精巧的灯塔模型，度过许许多多静谧的时光。有些守塔人还会做风筝，到灯塔的外廊去放飞，或者用海钓收获的鱼虾来调剂千篇一律的豆子晚餐。有些守塔人从陆地带来种子和泥土，利用礁石上的坑洼处来种菜。

守塔人所需的一切供给都要靠勤务船从大陆运来。勤务船每隔数月都会送来点亮塔灯需要的灯油、可以吃三四个月的食物（大多数是罐头和经过干燥处理的食物），以及饮用和盥洗用的淡水（收集的雨水从来都不够用）。勤务船还会带来一盒盒的铅笔和新袜子。它带来的流动图书馆，由许多装满书的小柜子组成，这样的小小图书馆在各处灯塔之间轮换。守塔人只能通过勤务船与外界联系。它带来陆地上的新闻和亲人的来信。对与世隔绝的灯塔居民来说，最重要的是勤务船会带来访客：一位医生或者地区督察官，甚至是守塔人的丈夫或者妻子。勤务船还会带走很多，例如退休的守塔人或者一封封家书。对许多人来说，寄出信件之后苦苦等待数月，期盼回信的焦虑是无法忍受的，所以他们把信件交给所信任的大海。曾经有位供职于英国尼德斯灯塔的守塔人，他将所有的信件封进漂流瓶，投入大海，估计只需五天就可以漂到陆地！

假如灯塔是建造在礁石或者悬崖上的，勤务船就无法靠近。于是船上的人就要划一艘小木船到小岛上岸。如果忽然遇到狂风大作，船上的人就用绳索将补给和访客从勤务船的甲板拉到灯塔上去。他们在绳索上吊一个帆布座椅，然后将补给放到上面。访客也是坐着这样的座椅到灯塔上的。他们在海浪上空摇摇晃晃，通常会被海水打得浑身湿透。

守塔人用生命守护着海上的每一艘船只。有很多守塔人冒着生命危险去营救沉船事故中的幸存者。许多守塔人为此牺牲了宝贵的生命。每座灯塔的背后，都有值得传颂的英雄事迹、惊心动魄的冒险故事以及平凡冗长的岁月。从20世纪20年代起，电灯逐渐取代油灯。很快，自动化的机器取代了发条装置，守塔人的职责就此结束。古老的灯塔变成了屹立不倒的标志，而不朽的故事会永远被后人传颂。

这本书里描绘的灯塔，位于我曾经住过的加拿大纽芬兰最北端的一座小岛。但是正如赫尔曼·梅尔维尔在小说《白鲸》里所写的："这个地方不在任何地图上，真实的地方从来都是这样。"

浪漫与严谨融为一炉的图画书杰作

阿甲

初读《你好灯塔》，让我忍不住想起电影《海王》，尽管侧重点不同，但都有灯塔看守人一家子的故事。有趣的是，它们都诞生于2018年，图画书获得了2019年凯迪克金奖，而电影则获得了极佳的票房收益。

看来，全世界的人们都非常喜欢了解灯塔看守人的生活故事。这样的故事乍一看浪漫极了：大海茫茫，一处礁石岛上矗立着一座美丽的灯塔，守塔人沉稳老练，恪尽职守，保证了周边航海导航的正常运行，而且还热心帮助海上遇险的人们；日升月落，潮起潮落，四季更迭，海上美景被守塔人一人独享，直到有一天他的伴侣来到，与世隔绝的孤独变成了二人世界的甜蜜，接着又迎来他们相爱的结晶，宁静、神秘又充满希望……看过电影《海王》，恐怕许多人会忍不住萌生去当个守塔人的强烈愿望，尤其羡慕那个守塔人居然跟海里的公主生下了一个海王级别的神奇儿子！

不过，你要是读过波兰作家显克微支的小说《灯塔看守人》，想法恐怕会完全不同。显克微支很坦白地交待了守塔人的日常工作，断定“灯塔看守人差不多就等于一个囚犯”，因为每天都被困在那里，按照规定做着重复刻板的工作，等着外面的人送粮食和淡水。守塔人的工作虽然算不上很艰苦，但非常孤独，他们过的“简直是一个隐居苦修者的生活”。小说中的守塔人是一位具有传奇色彩的七十岁老人，尽管他充满热情且勤勤恳恳，却仅仅因为一次疏忽没有点灯而被罢免。读罢这样的小说，估计是不会有人愿意从事这份职业的。

相比之下，如果你真的对灯塔好奇，想要对它有真实而全面的了解，这本《你好灯塔》可能是最佳的入门选择。这本图画书既有浪漫的一面，又有相当严谨的一面，而因为主要面向满心好奇的儿童读者，所以讲述起来趣味盎然。

故事是虚构的：一位新任守塔人到来，接着他的妻子跟随而来，然后他们的孩子诞生，最后新技术的引入导致不再需要守塔人，守塔人一家搬到海岸边，时常远眺他们曾日日夜夜坚守的灯塔。在虚构的故事线索中，确实不乏浪漫色彩：用漂流瓶寄出的情书，坐着吊篮来到灯塔的美丽妻子，冰面上玩耍的海狮与带宝宝游过来的抹香鲸，在圆形房间里徘徊等待新生儿的到来，在灯塔日志中记录下孩子的诞生……特别值得一提的是，这些看似虚构的浪漫情节都经过了考证，都曾经真的发生过。

作家兼插画家苏菲·布莱科尔不是第一次斩获凯迪克金奖，在2016年，她就因《寻找维尼》拿过金奖。尽管她不是那本书的文字作者，但在处理写实题材时，那种将浪漫与严谨融为一炉的本领给人留下了极为深刻的印象。

我曾在2014年夏天和家人一起去参观过缅因州的波特兰灯塔，那是一个很空旷的海边公园，灯塔本身还在运作，但已依赖电力完全实现了自动化，守塔人已成历史。那里有关于灯塔的各种介绍，最触动我的是一块金属牌匾上刻着曾经照顾过那座灯塔的每一位守塔人的名字和守护时间。每一个人背后都有一个故事，而守塔人都必须记录灯塔日志，所以他们的故事其实都有文字可考。那一次，我们在灯塔公园逛了半天，仍然意犹未尽。

据说，为了创作《你好灯塔》，苏菲去过北美大西洋边的许多灯塔。这是何等地下功夫！在某种程度上，她是在用这本书重塑一段北美的灯塔史。除书中的图画故事之外，在她的后记中，我们还能了解到更多的延伸知识。

尽管这只是一本给孩子的图画书，但文字和图画的细节都做过精心的考证。

作为插画家，苏菲聪明地选择了竖长型的开本，恰当且直观地展示了灯塔的整体样貌。为了展示灯塔的内部结构，她采用了剖面方式，读者可以清晰地看到灯塔内部的工作与生活环境：最底层的淡水系统，第一层的门厅，第二层的储物间，第三层的厨房与起居室，第四层的卧室，第五层的工作间，第六层（顶层）的导航灯。细心的读者在前面搞清楚灯塔内部的结构后，就能很清楚地知道接下来的室内活动在哪里、环境如何。苏菲

在展示灯塔内部活动时，常常采用圆形的画框，明明是平视，却总有某种从灯塔顶上俯视的效果。最有意思的是夫妇二人等待宝宝出生的对开页，摊开后完整的圆形正好在中间，人沿着边壁走动，竟有一种钟表指针的感觉，似乎能让人感受到那种等待的急切。插画家在这里显然并未完全写实。

《你好灯塔》的插画高明之处就在这里，尽管它相当详尽地呈现了灯塔的结构、运行原理和守塔人的生活，却同样能用柔和、优雅的水彩画展现浪漫的情怀。读者可以只将它当作“灯塔之歌”，也可以深入挖掘细节，了解电气化时代之前灯塔运行的原理和相关的灯塔演变史。我想，充满求知欲和好奇心同时又乐于不停想象的孩子们，一定能充分享受其中的。

2019年2月21日 写于北京

灯塔日志
美好牌缝纫线
400码